WITHDRAWN

THE EVOLUTION
OF THE
EMOTION-PROCESSING MIND

Robert Langs

THE EVOLUTION OF THE EMOTION-PROCESSING MIND

With an Introduction to Mental Darwinism

Robert Langs

Foreword by
James O. Raney

Introduction by
David Smith

INTERNATIONAL UNIVERSITIES PRESS, INC.
Madison   Connecticut

First published in 1996 by
H. Karnac (Books) Ltd.
58 Gloucester Road
London SW7 4QY

Copyright © 1996 by Robert Langs

INTERNATIONAL UNIVERSITIES PRESS and IUP (& design) ® are registered trademarks of International Universities Press, Inc.

All rights reserved. No part of this publication may be reproduced, stored in a retrieval system, or transmitted in any form or by any means, electronic, mechanical, photocopying, recording, or otherwise, without the prior permission of the publisher.

Library of Congress Cataloging-in-Publication Data

Langs, Robert, 1928–
   The evolution of the emotion-processing mind : with an introduction to mental Darwinism / Robert Langs ; foreword by James O. Raney ; introduction by David Smith.
      p.   cm.
   Includes bibliographical references (p.   ) and index.
   ISBN 0-8236-1775-0
   1. Psychoanalysis and evolution. 2. Emotions. 3. Genetic psychology. I. Title.
BF175.4.E86L36   1996
150.19′5—dc20
                                                                                  96-24500
                                                                                     CIP

Manufactured in the United States of America

CONTENTS

FOREWORD by James O. Raney — vii

INTRODUCTION by David Smith — xi

PART I
Darwin and Freud

1. Psychoanalysis and evolution — 3
2. Darwin and Freud — 19
3. A communicative framework for evaluating Freud — 26
4. Freud's evolutionary thinking — 34

PART II
Some biological principles for psychoanalysis

5. Hierarchies in psychoanalysis — 51
6. The principles of evolution — 63

7	The unit of selection	72
8	The architecture of the emotion-processing mind	78

PART III
Evolutionary scenarios for the emotion-processing mind

9	Setting the stage for an adaptationist programme	103
10	An evolutionary scenario: the early hominids	116
11	Another scenario: *Homo sapiens sapiens*	129
12	Solving the problems of death and violence	150
13	The evolution of frames, rules, and boundaries	160
14	Assessing the accomplishments of evolution	168

PART IV
The emotion-processing mind as a Darwin machine

15	Mental Darwinism	179
16	Selectionism and the therapeutic process	195

REFERENCES	207
INDEX	215

FOREWORD

James O. Raney, M.D.

When I heard Robert Langs speak for the first time about 18 years ago, his dedication to validation of psychoanalytic theory and methods impressed me as much as it does now. This was some years after my psychoanalytic training. I then invited myself to a few of his Friday morning seminars. As practitioner and theoretician he demanded as much from students as he did of his own science and of himself. His method revealed, in the microcosm of the clinical hour, immediate validation or refutation of the analyst's behaviour and interventions. This validation, though theoretically driven (as is all scientific observation), was founded exclusively in the patient's expressions (fantasies and stories). Almost immediately the analyst could witness the effect of his work, learn from it, and alter his hypotheses accordingly. This method was fine in theory, but in practice it created great demands on the analyst's personal narcissism and investment in theory. Personal analysis and continuation of professional development became a necessity, rather than some theoretical ideal. His viewing the patient as final arbiter connotes honour and respect in a deeply confirmed manner. Winnicott caught this in his assertion that the best supportive therapy is knowing that you are being well analysed.

Once enlightened by new truth or better theory, turning back is not possible. Langs's method, adhering closely to scientific method, continually demands change in basic psychoanalytic theory and practice. Many have simply ignored him completely, perhaps because many of the truths in his ideas are not debatable. The emphasis in Langs's work is basic to psychoanalysis and expands Freud's basic discoveries. He brings an object-relations point of view into a vital contemporary focus.

As I am first a biologist, adaptation has long interested me. Later, psychoanalysis, as I first learned it, seemed comparatively taxonomic and static. Langs's adaptational/interactive perspectives, on the other hand, seemed to generate vitality in each clinical interaction. With derivative inference and validation, which Langs has refined as his communicative approach to psychoanalysis, the scientific method applies in the microcosm of the psychoanalytic hour with convincing effect. Langs's work is always provocative. It is rejected too often as inflexible! According to one well-known psychoanalyst, it is too hard. I have used Langs's ideas, taught them, and argued them with my colleagues. They are hard and can only be used according to the capacity of each therapist. Directly and from observations of students and consultees, each so informed, psychoanalysis becomes an event with recognizable change not only in the patient, but in the analyst as well. The latter may be, as Harold Searles first noted, essential to the "cure" of both parties.

In this ambitious and at times difficult treatise Langs provides a satisfying missing link between the adaptational science of evolution and psychoanalytic psychology. Modern evolutionary theory is far more complex and sophisticated today than Darwinian survival of the fittest. Modern psychoanalytic theory is also much more advanced and complex than Freud's early theories. Perhaps these advances and complexity were necessary before the two theories could be compared. Psychoanalytic science will benefit from this association. If Langs has his way, human evolution will benefit from improvement of the function of the human spirit—the basic unit of which Langs refers to as the emotion-processing mind.

Because today's promotion of brief "therapies", quick fixes, and instant gratification is selfish, short-sighted, and destructive to human survival, the depth of psychoanalysis is needed more than ever. Serotonin manipulations, media or political self-interest will not cure the costly destruction of person, institutions, and community caused by character and personality disorders. For our society and, as Langs points out, our species, failures of growth, stability, maturity, and wisdom are individually and collectively devastating. These are best repaired by getting at the heart of the matter, at the disorders of the unconscious functioning mind. One of our best bets is psychoanalysis as a general psychology and as a clinical treatment.

Robert Langs is, as usual, ahead of his peers. In this work, he offers his most serious challenge to conventional psychoanalysis (if there is such an entity). He finds in a related science perspectives and riches that can enhance psychoanalysis. The science of evolution provides abundant foundation for the existence of an adaptive unconscious mental functioning. He also fields important verification of his own adaptational ideas. This book supports the position that psychoanalysis, as the most comprehensive human psychology, continues to be valid, alive, and ever-changing. Its unique access to unconscious mental functioning preserves psychoanalysis as the best vehicle for scientific discovery of the human condition. Langs's enthusiasm about the great potential for psychoanalytic science and practice is, even among the company of his many psychoanalytic detractors, the most reality-based optimism of all. Langs secures psychoanalysis as science well within the hierarchies of human sciences.

Psychoanalysis promotes true autonomy at the deepest levels. Autonomy is independence from destructive personal history and is a prerequisite to maturation and wisdom. Autonomy in the analyst means capacity to follow the patient. The Langsian model adds the emphasis of accountability and responsibility of the analyst. This is a model for accountability and responsibility in our lives, with our compatriots, and to our future generations. This is a hopeful book—read on.

INTRODUCTION

David Smith

Towards a post-Darwinian psychoanalysis

> Evolution is built on accidents, on chance events, on errors. The very thing that would lead an inert system to destruction becomes a source of novelty and complexity in a living system. An accident can be transformed into an innovation, an error into a success.
>
> François Jacob, *The Possible and the Actual*

It was almost one hundred and forty years ago that Darwin (1859) proposed his remarkable hypothesis of natural selection to explain how it was that the great variety of living organisms inhabiting this planet had come about.

Darwin's theory was a newcomer, and it had to contend with two powerful rivals: Creationism and Lamarckism. Creationism is a theological claim that the origin of species can be accounted for by divine *fiat*. Creationism remained virtually unchallenged from prehistoric times until the eighteenth century, which witnessed the birth of the sciences of geology and palaeontology.

Geology showed that the earth was much older than the biblical claim of roughly four thousand years. Palaeontology demonstrated not only that life on earth was much older than the biblical estimate, but also that biodiversity has increased over time, a fact contradicting the Creationist thesis.

Lamarckism was an evolutionary theory proposed by the biologist Jean Baptiste Lamarck (1744–1829). Larmarck proposed a theory of evolutionary change based on what he termed the "law of use and disuse". According to Lamarck, changing environmental circumstances compel animals to alter their behaviour. Those body parts and functions that are used less as a consequence of environmental change would, claimed Lamarck, atrophy, whilst those used more would be augmented. These changes would then be transmitted to the creatures' offspring. For Lamarck, the environment *instructed* creatures—it caused them to make adaptive efforts. This general approach is therefore called *instructionism*. The explanatory weaknesses of Lamarck's theory, combined with its incompatibility with the post-Darwinian science of genetics, ensured its replacement by Darwin's much more potent alternative.

Darwin's story of biological speciation was elegant in its simplicity. It ran like this. The offspring of any living thing are never completely identical. Each new organism is different in some way from its brothers and sisters, as well as from the offspring of other parents from the same species. Many of these differences are, for practical purposes, inconsequential: they have no effect on the way that an organism lives, reproduces, and dies. However, there are other differences that give an organism some edge over its rivals. Think of wild rabbits. In a litter of rabbits no two rabbits will be identical. Some of these differences will be trivial—such as minute variations in fur distribution. But think of the consequence of one member of a litter being born with a greater capacity for speed than its brothers and sisters. All things being equal, Speedy the rabbit will be less likely to be devoured by a fox than her brothers and sisters. Because Speedy will be likely to live longer than her brothers and sisters, she will—again, all things being equal—be more likely to produce more offspring than her brothers and sisters. When Speedy does reproduce, her offspring may share their mother's capacity for speed and will therefore tend to be more successful reproductively than the offspring of slower rabbits, and so on for generation upon generation. According to Darwin, the vast array of diverse biological species, including humankind itself, could be explained as having come into being through this simple, blind, random process, which he termed *natural* selection.

The principle of natural selection involves three moments. One begins with an initial stage of *variation*. This is followed by a process of *selection*. In the natural world, local environmental conditions "select" which variants are to survive. The selection process can be imagined as a kind of filter, which favours those creatures that are best able to make use of their environment. The third stage is the *reproduction* of those items that have been selected, which gives rise to a new stage of variation, and so on. It was only when the mechanics of genetic transmission were understood that Darwin's theory became fully plausible. It was then realized that random variations in genetic material (the *genotype*) produced variations amongst offspring (the *phenotype*), which, in turn, are reproductively transmitted by the phenotype. Darwin's theory is therefore *selectionistic* rather than instructionistic, as was Lamarck's.

Long after the "new synthesis" of Darwinism and genetics, which established natural selection as the motor of evolution, scientists and philosophers began to realize that the variation–selection–reproduction sequence that drives evolution is a powerful template for thinking about other life processes, thus giving birth to the concept of "universal Darwinism" (Plotkin, 1990). Burnet (1959) found that, contrary to earlier instructionist beliefs, the immune system conforms to selectionist principles, calling his approach the "clonal selection theory of acquired immunity". Earlier immunologists had assumed that lymphocytes "learned" to produce antibodies corresponding to invading antigens. Burnet found, to the contrary, that the body possesses a vast array of diverse lymphocytes, only some of which are selected to proliferate through contact with corresponding antigens. Gerald Edelman (1992) has argued that the brain itself develops through the environmental selection of neural connections, calling his approach "neural group selection". According to this thesis, only certain of the brain's many neural connections prove to be of any adaptive use. These connections are reinforced, whilst the unselected connections fall away. The philosopher Sir Karl Popper developed a theory of science itself based on selectionism: "evolutionary epistemology". In Popper's view, there is a clear evolutionary continuity between the life of an amoeba and the scientific work of an Einstein (Munz, 1985). Nature works through the creation of variants (organisms or ideas), which are then selected (for their survival or truth value) and selectively reproduced (as new organisms or new ideas). Edelman (1992) refers to selectionist "sciences of recognition", which investigate

> . . . the continual adaptive matching or fitting of elements in one physical domain to novelty occurring in elements of another, more or less independent physical domain, a matching that occurs without prior instruction. [p. 74]

Is psychoanalysis a Darwinian science of recognition? It is well known that Freud was a philosophical *naturalist*, in that he believed that the human mind is amenable to natural-scientific explanation. Freud was trained as a biologist, and as a student he did research bearing on what was then the "Darwinian controversy" (Ritvo, 1990; Sulloway, 1979). Freud often expressed admiration for Darwin, whose *Origin of Species* was published in the year when the three-year-old Sigismund (later "Sigmund") and his family settled in Vienna. In spite of his praise for Darwin, Freud never incorporated distinctively Darwinian concepts into his psychoanalytic work. The two main Darwinian concepts of adaptation and selection are almost completely unrepresented in Freud's theory of mind. In fact, Freud remained an outspoken Lamarckian until his death in 1939. Freud was not alone in this retrograde attitude, as continued allegiance to Lamarckian concepts persisted within the social sciences for some time.

In recent years the work of psychologists such as Donald (1991), Pinker (1994), and Tooby and Cosmides (1987, 1990b, 1992) incorporated Darwinism into psychological theory, thus giving rise to the vigorous sub-discipline of "evolutionary psychology". Psychology, the social sciences (Eldridge, 1992), and even philosophy (e.g. Millikan, 1984, 1993) have begun to learn from Darwin through applying selectionist concepts in their respective domains.

Strangely, psychoanalysis has been reluctant to put right Freud's erroneous neglect of Darwinian theory. It is only in recent years that psychoanalytic writers such as Badcock (1986, 1990b, 1994), Nesse (1990b; Nesse and Lloyd, 1992), and Slavin and Kriegman (1992) have attempted a strong neo-Darwinian reconceptualization of psychoanalytic theory and technique, and it must be said that the suggestions offered by these writers have not been widely taken up or debated within the field. Ironically, psychoanalysis seems happy to leave biology behind.

In 1991 I gave an address to what is now the International Society for Communicative Psychotherapy and Psychoanalysis, calling for an evolutionary approach to the phenomena identified by communicative psychoanalysis (Smith, 1991). By virtue of its strong emphasis on adaptation and its implicitly selectionist approach to mental processes, the communicative approach seemed to be—at least potentially—a "science of recognition" in Edelman's sense. Robert Langs was present on that occasion and, with characteristic intellectual alacrity, brilliantly developed my rather vague, programmatic recommendations into a number of specific propositions about psychoanalysis in general and the communicative approach in particular. The present work testifies to the fruitfulness of that investigation. *The Evolution of the Emotion-Processing Mind* represents a decisive step forward in the "Darwinization" of psychoanalysis.

Going against the current of the general drift of psychoanalysis in the direction of hermeneutics, Langs has been adamant that psychoanalysis should define itself as a natural scientific discipline. As such, it must situate itself within the nested hierarchy of the sciences, clarify its relationship with adjacent disciplines, and look to more highly developed sciences for support and guidance. As such, Langs stands in opposition to those who regard psychoanalysis as a fully autonomous discipline, unconstrained by the broad logical and methodological norms of science. For Langs, like the philosopher Millikan, psychology—as the study of evolved mental functions—is a sub-discipline of biology. Even if one rejects the idea that psychoanalysis is properly a component of psychology, surely there can be no argument that psychoanalysis strives to understand human life. As such it must pay heed to biology, the science of life. *The*

Evolution of the Emotion-Processing Mind is a highly original synthesis of psychoanalytic and neo-Darwinian thought, providing a pathway to a more profound understanding of human nature, human suffering, and the role of psychotherapy in the amelioration of suffering.

PART I

DARWIN AND FREUD

CHAPTER ONE

Psychoanalysis and evolution

Modern-day perspectives make it clear that psychoanalysis can no longer be considered as anything but a biological science (Plotkin, 1994; see also Slavin & Kriegman, 1992). As members of the species *Homo sapiens sapiens*, we humans are biological creatures, and every aspect of how we function and adapt belongs to the biological realm. This includes, of course, the operation not only of our *brains*, but of our *minds* as well. And it is the human mind, the mind of *Homo sapiens sapiens*, that is the subject-matter of the biological subscience known as *human psychology*, of which *psychoanalysis* is a further division (Kitcher, 1992).

Viewing psychoanalysis as a biological science implies that biology and its many subsciences must of necessity be intimately connected with, and have the potential to deeply inform and bring fresh thinking to, the theory and practice of both psychoanalysis and psychotherapy (terms I will use interchangeably). In part this potential arises because the appeal to biology entails a turning to interdisciplinary research and thinking, a highly difficult but unusually rewarding undertaking (Kitcher, 1992).

Trained as a biological scientist, Sigmund Freud, the founder of psychoanalysis, initially had a strong commitment to this kind of pursuit (Freud, 1950 [1895]; Sulloway, 1979). He tried in a variety of ways to establish and sustain a biological foundation for the theory of psychoanalysis, though less so for its practice. Unfortunately, given the state of the available sciences at the time of his researches and writings and his own rather striking intransigence and failure to revise his position when the basic concepts of these sciences changed dramatically, Freud was never able to provide psychoanalysis with the interconnections with fellow sciences that he so arduously strove to establish (Kitcher, 1992).

Freud's lack of success and the growing complexity of the biological sciences have, with few exceptions, discouraged psychoanalysts from interdisciplinary pursuits. Nevertheless, if psychoanalysis is to flourish and evolve, it must by all means become an interdisciplinary science. Links to developmental psychology, neuroscience, and evolutionary biology are currently being forged (see, for example, Slavin & Kriegman, 1992), and they have already provided us with glimmers of fresh ways of viewing the human mind and human relationships in psychoanalytic terms. However, the ultimate outcome of these efforts will greatly depend on the status of the field of psychoanalysis itself, which must prove capable of forging a solid scientific foundation for its own theoretical constructs in ways that will enable its propositions to link up with other sciences in meaningful fashion, and to the enlightenment of both sides of these partnerships.

For some years now, the *communicative approach* to psychoanalysis and psychotherapy that informs this book has been striving to move in this direction (see especially Langs, 1992c, 1995a; Langs & Badalamenti, 1994a, 1994b, in press). Efforts have been made to expand the classical theory of psychoanalysis—which is almost entirely concentrated on the psychodynamics and personal genetics* of intrapsychic conflict, object

---

* I will use the terms *genetics* and *biological genetics* to refer to the genes contained within chromosomes, while the terms *personal, psychological, or historical genetics* will be used to refer to early life experiences as they relate to an individual's personal history.

relations, and self—by introducing a series of added dimensions to its basic propositions. These additional viewpoints include such complementary fields as formal science, systems theory, and psychoanatomy or model making—the pursuit of the architecture of *the emotion-processing mind, the mental module of conscious and unconscious functions that is responsible for processing and adapting to emotionally charged information and meaning.*

The present volume further extends these efforts by turning to the science of *evolution* in order to provide psychoanalysis with a basic yet broad set of evolutionary principles and perspectives for its theoretical structure and clinical precepts. The plan is to develop a series of fresh propositions and insights derived from *evolutionary biology* and *evolutionary psychology* in order to enlarge the evolutionary approach to psychoanalysis—the newly developing field of *evolutionary psychoanalysis.*

While by no means the first effort in this direction (see, for example, Badcock, 1986, 1990a, 1990b, 1994; Glantz & Pearce, 1989; Lloyd, 1990; Nesse, 1990a, 1990b; Nesse & Lloyd, 1992; Slavin & Kriegman, 1992), the present approach to uniting psychoanalysis and evolution has many distinctive features. Chief among them is the choice of a specific and fundamental *unit of selection* (Lewontin, 1970, 1979; Tooby & Cosmides, 1990b) for the study of evolutionary change and emotional adaptation, namely, the aforementioned *emotion-processing mind*, which will be the focal point for my evolutionary investigations. With the aid of clinical observation and communicative thinking on one side, and recent advances in Darwinian theory on the other, my intention is to develop two original presentations designed to clarify the structure and operations of the human psyche and, by that means, to illuminate freshly both psychoanalytic theory and the therapeutic process.

1. For the first of these offerings, I turn to the six-million-year history of the hominid line and propose a scenario, a so-called *adaptationist programme* (Dennett, 1995; Gould, 1987; Gould & Lewontin, 1979; Kitcher, 1985; Mayr, 1983, 1983; Nesse & Williams, 1994; Slavin & Kriegman, 1992; Tooby & Cosmides, 1990b), for the historical unfolding—the evolutionary development—of the emotion-processing mind.

2. The second offering pertains to the immediate adaptive efforts of this mental module and introduces the concept of *mental Darwinism*—a view of the emotion-processing mind as a Darwin machine (Calvin, 1987; Plotkin, 1994) that operates according to the universal principles of selectionism and Darwinian evolution (Dawkins, 1983).

These two means of bringing evolutionary principles to psychoanalysis—one a historical perspective, the other a contemporaneous one—are at the heart of this book.

Why evolution?

It is widely acknowledged that the Darwinian theory of evolution, including its neo-Darwinian extensions and revisions, is *the fundamental theory of biology* (Dennett, 1995; Eldredge, 1995; Plotkin, 1994). This claim is made largely because *adaptation and speciation* appear to be the two most distinctive aspects of biological organisms (Dawkins, 1976b, 1987; Gould, 1982; Mayr, 1974; Plotkin, 1994; Ridley, 1985; Slavin & Kriegman, 1992). In addition, the readiness with which the principles of Darwinian theory generalize from evolution to other biological and adaptive processes speaks for the remarkable power and cogency of evolutionary constructs (Dennett, 1995; Plotkin, 1994). Most certainly, psychoanalysis must reckon with and incorporate the theory of evolution into it own theoretical edifice.

Darwinian theory has two fundamental components:

1. The study of immediate biological adaptations as they exist within a given species and, more narrowly, in a particular individual organism.
2. The study of the origins, history, forces, and means by which these adaptations have evolved. This includes exploring how these coping strategies and mechanisms have dealt with and solved or failed to solve specific environmental challenges, and how they have found form and function over exceedingly long periods of time. As stressed by Tooby and Cosmides (1990b), this pursuit is the essence of Darwinian evolutionary science.

Evolutionary theory and psychoanalysis

We must ask, then, of what particular relevance is evolutionary theory to psychoanalysis and psychotherapy? In principle, the answer is unmistakable: as the embodiment of the fundamental postulates of biology, evolutionary theory and thinking must *per force* be critical as both an *influence on, and constraint for, psychoanalytic theory and practice*. That is, evolutionary principles must in some form apply to and inform the entire psychoanalytic domain. But in addition, as the more fundamental of the two theories, Darwinian principles and insights set limits on psychoanalytic propositions—*no psychoanalytic concept or precept can contradict established evolutionary principles*. Indeed, it is one of the main goals of this book to demonstrate the enormous importance of this influence and constraint, and the surprising extent to which Darwinian theory clarifies and calls for a reshaping of psychoanalytic propositions and clinical techniques.

The ramifications of Darwinian theory are rich and intricate; a brief listing will serve to introduce us to its potential as a resource for the fields of psychotherapy and psychoanalysis and outline the pathways to be explored in this book. Evolutionary theory pertains to and interacts with psychoanalysis in at least the following ways:

1. The overarching position of Darwinian theory compels us to recognize that the investigation of emotionally relevant *adaptations* must be a central, if not the single most vital, feature of psychoanalytic theory and research. In terms of current parlance, this implies that *the adaptive meta-psychological viewpoint* of psychoanalysis (Gill & Rapaport, 1959; Hartmann, 1939) must stand at the forefront of its theoretical propositions and clinical perspectives.
2. Emotionally relevant adaptations, by definition, are carried out by—are the functions of—an entity or *mental module* of the human mind (Cosmides & Tooby, 1992; Gazzaniga, 1992; Tooby & Cosmides, 1987, 1990a, 1990b, 1992) called *the emotion-processing mind*. Although this adaptive organ is an expression of underlying *brain* mechanisms (Edelman,

1987, 1992; Restak, 1994), psychoanalysis, as a branch of psychology, must operate in the *mental domain*. That is, the province of psychoanalysis is the *human mind* and the *emotion-processing mind* in particular. Thus, this mental module—the emotion-processing mind—must be explored within the mental realm and viewed as an entity whose structure, manifestations, intakes and outputs, general properties, and adaptive functions are distinct from those involving the human *brain*—however else the two domains may interact and be related.

a. The psychological observables and data of psychoanalysis must include not only affective states and subjective thoughts, but also mental outputs like *language and communication*.

b. It follows, too, that the basic problems, postulates, explanations, models, and predictions of psychoanalysis must initially be defined in terms of mental phenomena and adaptive resources.

c. The psychoanalytic study of *adaptation* can be defined as the investigation of mentally grounded coping responses to environmental impingements—emotionally charged external events that are fraught with information and meaning (internal events are, as a rule, secondary evocative stimuli). These adaptations are reflected mainly in affects, thoughts, bodily responses, language-communication, and behaviour.

d. A basic axiom of the psychoanalytic position states that emotionally relevant adaptations occur on two levels of organization and mentation. One level is attached to awareness and essentially operates consciously or with ready access to awareness, while the other level has no direct connection to awareness and operates unconsciously. Indeed, the postulate of *unconscious processing or efforts at adaptation* is a hallmark of psychoanalytic theory.

e. Given that *the emotion-processing mind is the basic unit and means of emotionally related adaptation*, the investigation of its evolved, universal, deep structure and the nature of its adaptations appears to bring us to a fundamental level of evolutionary investigation. Probes into more complex configurations such as overt behaviour, relatedness, interaction, and the like can and must build on this foundation.

f. For humans, however, knowledge acquisition and effective adaptation require secured settings and sound relationships—both learning and being able to develop sound coping skills depend on relating. Thus, *communication, adaptation, and relating* are each, in some sense, fundamental to human existence and survival.

g. Nevertheless, all three modalities rely on and function on the basis of the operations of the emotion-processing mind, which appears to be an essential foundation for the various means by which humans adapt to their environments. It follows, then, that there are distinct advantages to making evolutionary explorations of communication and the processing of information and meaning a top priority, because they can reveal most clearly the conscious and unconscious capabilities and properties of the emotion-processing mind.

3. The structural architecture and adaptive functions of the emotion-processing mind have evolved as part of the history of the biological species that have occupied this planet. This means that the psychoanalytic investigative agenda should, of necessity, include the development of a workable, testable, and potentially falsifiable or refutable scenario of this history as it reveals and clarifies the forces that shaped the present design of this mental module— its many assets and its surprisingly large number of liabilities.

a. In the science of evolution, this type of scenario is called an *adaptationist programme*. In principle, it must be developed according to the rules of the scientific method and in light of both psychoanalytic and Darwinian principles and insights. The scenario must be as unbiased as possible, conform to all available data, yield insights into the nature of specific adaptations and the environmental issues that they were designed to solve, present perspectives that are unavailable through any other means, and be rechecked repeatedly because of the ease with which errors can be made in this kind of endeavour (Gould & Lewontin, 1979; see also Dennett, 1995; Lewontin, 1979; Mayr, 1983; Slavin & Kriegman, 1992; Tooby & Cosmides, 1990b). We can have confidence in an adaptationist programme only when there is a confluence of such factors as supportive observations, special explanatory powers, and the validation of predictions and hypotheses based on the programme. These are some of the essential safeguards against the high risk factor in these very necessary and creative efforts. Indeed, despite the risks, this kind of undertaking can provide insights into the human mind and its adaptations that cannot be achieved by any other means.

b. On the psychoanalytic side, the first requisite for an adaptationist programme is a sound grasp of the architecture of the emotion-processing mind, without which a fundamental evolutionary tale cannot be constructed. Beyond that, much will depend on a particular author's means of observing nature (and the therapeutic interaction in particular), of organizing and giving meaning to these observations, and the version of psychoanalytic theory he or she turns to for guiding principles.

c. On the evolutionary side, the Darwinian principles of evolution must be properly defined and then used to guide and constrain the unfolding psychoanalytic scenario. We must tailor the basic concepts of evolution to serve the psychoanalytic domain. Concepts such as random variation, adaptive competition, selective pressures, and natural selection must be brought to bear on the differential reproduction of those emotionally relevant mental structures and adaptations that seem most to enhance survival and reproductive fitness.

   i. Of special importance in this regard is an understanding of the evolved design of the emotion-processing mind in light of the shaping powers of the *selective pressures* that have arisen from its ever-changing *environments—a comprehensive term that includes both the social and physical, the animate and inanimate domains, and related events.*

ii. Another crucial Darwinian concept is the *principle of natural selection or selectionism.* Natural selection should be thought of as a passive, mindless set of principles, as an algorithm (Dennett, 1995) through which various competing adaptations are tested and those that are most viable are chosen for favoured reproduction.

iii. In the neo-Darwinian literature, this principle of *selectionism* is contrasted with the Lamarckian principle of *instructionism.* For the study of evolution, the issue is that of genetic transmission versus the transmission of acquired and learned responses, while for the study of immediate adaptations, the problem lies with choosing between a process of environmental *selection* from an organism's available repertoire of adaptive responses versus environmental *instruction* to the organism as to how to react or adapt to external impingements.

iv. At present, evolutionists favour the idea that *selectionism* is the mode of operation for the most vital adaptive systems of *Homo sapiens sapiens,* including the *immune system,* intelligence, cognition, and the *brain* (Edelman, 1987, 1992; Gazzaniga, 1992; Jerne, 1955; Nesse & Williams, 1994; Plotkin, 1994; Tooby & Cosmides, 1987, 1990a, 1990b, 1992). Hypothetically, then, we would expect the emotion-processing mind also to follow this principle in its adaptive responses (see part IV).

4. Evolutionary theory states that current adaptations have both *proximal or immediate* and *distal or evolutionary/historical* causes (Slavin & Kriegman, 1992). The definition of both types of causality is essential for a comprehensive picture of emotional health, psychopathology, and the dynamics of the therapy process, although the problem of distal causes is the essence of evolutionary theory (Tooby & Cosmides, 1990b).

   a. In this regard, Darwinian theory indicates that the long-standing emphasis in psychoanalysis on proximal causes presents only a minor part of the picture, whether it pertains to emotional adaptation or to the therapeutic process. Evolutionary psychoanalysis has been created to account for the other half of these pictures—distal causes. As a distinctive viewpoint based on particular types of observations of a kind that is essential to evolutionary insights, this endeavour inevitably should bring fresh perspectives to current analytic thinking.

5. One of the most powerful theorems of Darwinian theory states that the *environment*—comprehensively defined as *the fitness landscape or environment, the adaptive environment, or the environment of evolutionary adaptation* (Dennett, 1995; Eigen, 1992; Tooby & Cosmides, 1990b; Wright, 1931, 1932)—is the *primary* source of the stimuli that evoke adaptive responses in an organism. Furthermore, the environment plays a central role as a causative factor in, and source of, the *selection pressures* that shape the evolutionary histories of organisms.

   a. This principle, which pertains to all living and extinct species, implies that *external trigger events or impingements* on human organisms and their emotion-processing minds—stimuli emanating from an individual's social and physical environments—are the *primary* causes of both adaptive and maladaptive or dysfunctional responses.

   b. Because humans have extraordinary, yet remarkably limited, powers of self-awareness (Langs, 1995a; Ornstein, 1991), inner mental happenings such as affects, thoughts, and fantasies do have some power as *secondary* adaptive issues or triggers for a given individual and his or her emotion-processing mind, especially when they pertain to harm- and death-related issues. In addition, physical/bodily stimuli also operate as internally derived adaptive issues, mainly when they involve life-threatening injuries and illnesses.

   It is well to realize, however, that many of these inner experiences are themselves, first and foremost, adaptive reactions to environmental impingements, and that only secondarily do they operate as personally evocative triggers.

   c. In addition, for *Homo sapiens sapiens*, selection pressures may arise from a variety of other sources—e.g.

personal inventions (Baldwin, 1896), culture, ecological advances, and the ecosystem (Eldredge, 1995). A broad view of selection pressures is essential for a comprehensive theory of evolution.

6. Evolutionary theory also impacts on psychoanalysis by compelling us to take a fresh look at the causative factors in emotional health and illness, whatever the form of dysfunction—intrapsychic or interpersonal. We must be prepared to revise or eliminate all psychoanalytic precepts and beliefs, however cherished, in light of evolutionary principles.

    a. For example, the primacy that psychoanalysis has afforded to intrapsychic fantasies as causal factors in psychopathology must be downgraded to a secondary role in light of the evolutionary principle that environmental factors are the central issue for all of the basic adaptive and maladaptive responses of an organism. Similar considerations apply to the recent stress on personal interpretation when relational experiences and conflicts are formulated (Slavin & Kriegman, 1992)—external reality must be afforded its full due.

7. Evolutionary theory calls forth two polarities of considerable relevance to psychoanalysis:

    a. The first of these places the inherited/genetic causative factors of behaviour, broadly defined to embrace all organismic responses, at one end of a continuum, and all of the other non-genetic, non-instinctual factors that affect behaviour at the other end. The latter include embryonic and post-partum developmental circumstances, as well as life experiences, including accidental traumas and positively toned events. As with all *nature–nurture* considerations, it is well to understand that these two fundamental factors are always interrelated and that both are ever-present in shaping an adaptive response, even though one or the other may exert the greater force for a given individual at a given moment.

    b. The second polarity is far less familiar to psychoanalysts. It places the *average expectable environment* at one end of the continuum and *unexpected futures* at the other.

The average expectable environment concept, developed as a framework for ego development and the need for environmental support, was a notable concern of Hartmann (1939) and of Hartmann and others in the 1950s (Hartmann, Kris, & Lowenstein, 1951) and was part of their adaptive viewpoint—an extension of Freud's rather global ideas in this area. This concept has had some elaboration in the recent writings of evolutionary psychoanalysts who have used it to frame the seemingly genetically determined issues and conflicts that arise in parent–child and patient–therapist interactions (Slavin & Kriegman, 1992).

At the other end of this continuum lies *the unexpected environment or uncertain futures problem* (Plotkin, 1994; Waddington, 1969). While the average expected environment concept points to the need to stabilize the environment and parental and therapeutic care and stresses the genetic basis of adaptations, the unexpected environment problem points to ways in which the environment inevitably becomes unstable—often to an extreme degree. The focus therefore shifts from how organisms are genetically equipped to deal with the familiar and expected environment to how they are designed to adapt to unfamiliar and unprecedented environmental impingements.

Because genetically determined resources are very slow to change and therefore cannot readily respond to immediate and short-term environmental alterations, resources like learning and intelligence play a major role in solving this latter type of problem. As interest in sexual selection is extended to problems of economics and survival, the means by which we adapt to unprecedented experiences and events becomes a critical concern. These issues are of considerable relevance to our study of emotional cognition and to psychoanalysis in general.

8. Evolutionary considerations are one way of bringing into focus and freshly clarifying still another important polarity of human nature—*individuality versus universality*. By and large, psychoanalysis has stressed the distinctive and individual features of each patient and therapist, and the treatment experience they create together. In the extreme, many psychoanalysts have argued against the possibility of laws of

the mind, regularities of the therapeutic interaction, or significantly meaningful human universals as they pertain to our field.

Short of this extreme position, many psychoanalysts have adopted a *weak position in regard to selected universal attributes.* For example, they assume that so-called transference responses by therapy patients and countertransference responses in therapists are universal phenomena, as are the psychosexual stages of development and interpersonal psychological configurations like the Oedipus complex. They also afford a vague sense of the universality to id, ego, and superego development, structure, and functioning. Even so, psychoanalysts in general are inclined to emphasize the individuality of the expression of these entities far more than their shared features.

Darwinian theory brings a fresh perspective to this problem of individuality versus universality. The theory of evolution articulates general principles of evolutionary change that apply to all living creatures, including humans. Evolutionary theory thereby pertains to individual variation as it develops within the constraints set by generally lawful and universal rules of evolution such as the selection process. The theory also speaks for regularities and consistencies that relate to the interaction between genes and their expression in the formation of the human mind and body (the effects of genotype on phenotype).

Evolutionary theory therefore compels psychoanalysis to discover and explore its universals to complement its present stress on individuality. This calls for a study of the universal attributes of such fundamental structures as the emotion-processing mind and other basic mental entities and functions and asks us to recognize that these general properties constrain individual predilections. This viewpoint has been developed in some detail by the communicative approach and by evolutionary psychoanalysts and others who have studied relational and other phenomena from the evolutionary vantage point (Brown, 1991; Langs, 1995a; Nesse, 1990b; Slavin & Kriegman, 1992; Tooby & Cosmides, 1990a).

a. This contribution from evolutionary biology both supplements and supports findings from recent, mathematically grounded, communicative research efforts, which have revealed deep laws and regularities of human communication and of the human mind (Langs, 1992c; Langs & Badalamenti, 1992a, 1992b, 1994a, 1994b, in press). In addition, efforts in the new field of *psychoanatomy*—exploring and defining the architecture of the emotion-processing mind (Langs, 1986, 1987a, 1987b, 1988, 1992b, 1995a)—also have revealed the existence of a universal design with individual variations that are constrained by its general features.

Individuality constrained and guided by universals laws and regularities is a critical and ever-present evolutionary principle of great import that needs to be fully embraced and explored by today's psychoanalysts.

9. Evolutionary considerations also introduce a wide range of additional issues and perspectives into psychoanalytic thinking.

   a. As a science of the emotional domain, the evolutionary history of psychoanalysis itself needs to be mapped (Sulloway, 1979). This effort should include a study of the progressive and regressive aspects of its evolutionary trajectory and the identification of the positive and negative forces and the specific, unique selection pressures that account for its historical development.

   b. The *isomorphisms or fractals—the self-replicating patterns of nature across entities and species*—also need clarification. These repetitive configurations have a bearing on the fundamental adaptive mechanisms and solutions to environmental challenges that natural selection has invoked at different junctures in the evolution of biological species. Understanding these homologies can shed light on many aspects of human emotional adaptation and other psychoanalytic phenomena, including the evolutionary history and present status of the emotion-processing mind (Kauffman, 1995; see also chapters 10 and 11).

   c. Counterbalancing these shared attributes is a set of *emergent and essentially unprecedented features* that are unique to *Homo sapiens sapiens* (Bickerton, 1995), including their emotion-processing minds. They too need to be identified and examined.

d. Biological organisms and their adaptations are structured *hierarchically*, as are the forces that influence them (Dawkins, 1976a; Eldredge, 1995; Eldredge & Grene, 1992; Eldredge & Salthe, 1984; Grene, 1987; Plotkin, 1994; Salthe, 1985). To be comprehensive sciences, then, both evolutionary theory and psychoanalysis must themselves have hierarchical organizations. The mapping of their hierarchies is another fresh task for our field (see chapters 5 and 6).

The goals of this book

To conclude this broad introduction, I will list the main goals of this book. They pertain both to the evolutionary history of the emotion-processing mind and to its immediate modes of adaptation. They are:

1. To offer a new approach to the unification of the theories of evolution and psychoanalysis.
   a. To indicate how this union illuminates psychoanalytic theory and the clinical practices of today's psychoanalysts and psychotherapists.
2. To define the adaptive capabilities of the emotion-processing mind, perhaps the most fundamental unit of human adaptation in the realm of emotional experience, and to do so in light of its evolutionary history.
3. To delineate an adaptationist programme that accounts for the six-million-year evolution of the emotion-processing mind and to clarify thereby some of its most puzzling current features.
4. To provide support for the contention that emotionally charged behaviours, and their underlying mental structures and processes, have evolved in a manner that is comparable to that of morphological (physical) structures like limbs and brains.

5. To advance the concept of *mental Darwinism*—namely, that significant aspects of the current adaptive operations of the emotion-processing mind are governed by Darwinian principles of selection and evolution—and to explore the consequences of this new proposition.

The pursuit of evolutionary scenarios for the human *brain* is a daunting effort—brains do not fossilize and skull markings and other ingeniously discovered indirect clues are all that is available for such efforts. Comprehending the evolution of cognitive functions—non-emotional ways of knowing and adapting to the world, including the use of language—has a host of problems of its own (Bickerton, 1990, 1955; Donald, 1991; Tooby & Cosmides, 1990b). We can therefore expect that charting the evolutionary course of the emotion-processing mind will be a difficult and arduous task. Still, the very uniqueness and uncertainty of this challenge makes evolutionary psychoanalysis an especially promising and exciting undertaking.

CHAPTER TWO

Darwin and Freud

The two giants of biological science of the nineteenth and early twentieth centuries most certainly were Charles Darwin and Sigmund Freud. Freud was but three years old when Darwin, after much delay, published his *Origin of Species* (Darwin, 1859). Although Darwin died in 1882, long before Freud's theories were fashioned, Freud worked with or knew biologists like Haeckel and Claus who were staunch supporters of Darwin's then controversial theory of the evolution of species. Freud therefore knew a great deal about Darwin's work (Badcock, 1994; Kitcher, 1992; Ritvo, 1990; Sulloway, 1979).

Although they never met in peace as collaborators or on the battlefield of competing sciences, where validation and falsification are the ultimate arbiters, we can by way of introduction establish which theory and theorist adopted the more compelling and fundamental position. On that basis, we can define the place of each science in the hierarchy of biological sciences and use this perspective to examine Freud's interaction with Darwinian theory.

For a host of reasons—for example, the way each man worked, observed, and reasoned; the nature of their respective fields of primary endeavour; and the range of their purviews and vision—Darwin had many advantages over Freud as a researcher and scientist. Focused as it is on all of biological nature, Darwin's theory of evolution via natural selection clearly stands at a more fundamental level than Freud's theory of psychoanalysis (and any later versions of this theory as well). Indeed, in a recent book on evolution, Dennett (1995), in asking who will stand as the greatest genius in the history of science, cited Newton, Einstein, and Darwin as the sole contenders, and selected Darwin hands down.

The crucial implication for us here has already been stated: the theory of evolution is the fundamental theory of biology, and psychoanalytic theory is at a lower point in the hierarchy of biological sciences and is nested within evolutionary theory. According to the rules of hierarchical structure, then, psychoanalysis cannot entertain a hypothesis that violates Darwin's established theory of evolution.

There is, however, a broader point: Darwin's theory reigns supreme because it generalizes well beyond the explanation of the evolution of biological species; it is a theory that applies in some fashion to an enormously wide range of natural phenomena (Dennett, 1995; Plotkin, 1994). Darwinian principles are relevant to the operations of such diverse human entities and capacities as the brain, the immune system, cognitive functioning, and the use of language, and they are a factor in a wide range of additional specific modes of adaptation carried out by a vast array of other biological species. These principles also apply to the development of populations, cultures, and the biosphere, and to the pursuit of science and other collective and individual human endeavours (Burnett, 1959; Eldredge, 1995; Gazzaniga, 1992; Jerne, 1955, 1967; Pinker, 1994; Pinker & Bloom, 1990).

There is, however, evidence, some of it reflected in this book, that just as the study of speciation led to a set of widely applicable universal principles of biological organization and functioning, Freud's study of neurosogenesis and of the transactions of the psychoanalytic situation also could have produced similar universal and broadly applicable insights and principles. Dennett (1995) calls this "starting in the middle", which alludes to how a scientist can initiate investigations by focusing on a narrow set of phenomena and subsequently generalize to fundamental concepts and basic historical truths. Properly framed and conducted, Freud's investigations of emotional adaptations could have yielded insights that generalize to all of biology—and they can still do so to this day.

Comparing Darwin and Freud

Why then was Freud unable to produce this kind of universal theory? There are many answers to this question. One of these may be that Darwin was a meticulous observer and Freud an imprecise one. Then, too, Darwin was definitively searching for *universal principles* applicable to all of biological nature, whereas Freud was more concerned with the narrow realm of the emotional domain and with *individuality* as it related to emotional development and the analytic interaction—a particular patient's neurosis and a given analyst's countertransferences. While Freud also sought some rather broad universals (see chapter 1), this issue was not a prominent theme in his writings and thinking. In addition, he abandoned the promising pursuit of the basic design of the psychic apparatus, as he termed it, which he had initiated early in his career (Freud, 1900a, 1950 [1895]) and which might have kept him focused on the biological foundation of emotional adaptations.

Furthermore, while Darwin sought and discovered a set of well-honed fundamentals, Freud forged a theory with few, if any, fundamentals in the sense of *basic entities*, functions, laws, regularities, and structures. Instead, he fashioned a theory with high-level, global inventions like transference, countertransference, resistance, infantile sexuality, and vaguely conceptualized notions of an unconscious mind and unconscious meaning—a mixture of entities and processes that were all but impossible to define empirically or to deconstruct into basic elements. His biological principles, such as the reality and pleasure principles and the principle of constancy, as well as the life and death instincts, were poorly documented and open to serious challenge; also, they tended to violate the basic principles of evolution, which stress adaptation, environmental evocations of behaviour, and the primary quest of all organisms for survival and reproductive success.

Freud's theory was hermeneutic and concept-driven, a search of nature for confirmatory evidence for theoretical constructs rather than an observation-driven theory that could readily be modified by inexplicable observables—a basic reason why Freudian theory is not falsifiable. In evolutionary terms, this type of theory is *instructionistic* (Plotkin, 1994) in that it directs the environment and nature to produce confirmatory evidence for an existing theoretical postulate and screens out data and phenomena that contradict or cannot be explained by the theory. As with all instructionism, the method is limiting and has little potential for creative discovery and change.

In contrast, Darwin was a meticulous observer whose methods were *selectionistic* in that his observations directed a selection among competing theories for the set of ideas that best fit with his observables. Darwin struggled to integrate a diversity of natural observations into a theory that could account for these phenomena and unify them in some fashion. He worked with observables that could be readily verified by others, while Freud worked with observables that were uncertain and contaminated by theoretical constructs that tended to be intermixed with the phenomena he explored. As a result, Freud's ideas were idiosyncratic and difficult to verify, and his theory was offered on a level that proved impossible to extend to other domains.

Obstacles for Darwin and Freud

Still another aspect of the respective work of Darwin and Freud played a notable role in the differences in the structure and fate of the theories that each formulated. Darwin's domain was the wonderful world of natural species, past and present. Although the extinction of species was part of his scenario, his primary thrust was towards life and living, survival and reproductive success. His knowledge base essentially involved neutral or even attractive entities, and the insights he was seeking and the ideas that he developed were primarily intellectual and strikingly enlightening.

Darwin's main obstacles were the egocentricities and strongly anti-scientific religious leanings of his contemporaries, and their view of themselves as a species apart from the rest of nature. They maintained a powerful belief in creationism—that a design must have a designer and that only God has the power to design the wonders of the biological world (see Dennett, 1995, for further details). Darwin had to overcome human and personal prejudices, but he was able to delve into his data with such intensity and to muster so much evidence and such a compelling vision—aided, as is well known, by the writings of Malthus on the issue of competition for supplies of energy among organismic populations—that he came through it all with the basic principles of evolution in hand, a truly revolutionary theory that changed our picture of ourselves, nature, and the world around us.

What, then, were Freud's obstacles to developing a sound and revolutionary theory? The domain of psychoanalysis is restricted to a particular aspect of nature—the vicissitudes of human emotions and emotional life, and related psychological dysfunctions. While emotions have a long history in animal life and behaviour, there are many aspects of emotional conflict and adaptation that, due mainly to language acquisition, are almost exclusively human. Freud's definition of psychoanalysis reflects these emergent and distinctive features by marking the field as the investigation of emotional life *as illuminated by unconscious psychodynamic processes and elements*; dynamically unconscious processes appear to be a uniquely human attribute.

Essentially, then, Freud's self-assigned task was to define and explore the *unconscious* realm of human experience as revealed through human communications and behaviours. However—and this is a key point—the unconscious realm of experience harbours impingements and meanings, perceptions of others and ourselves, that have been banished from awareness because they are terrifying, conflicted, the cause of dysphoric affects, and essentially far too dangerous and potentially disruptive to articulate, expose, and behold consciously. While Darwin's obstacles involved prejudice, Freud's barriers involved the dread of emotionally charged unconscious meaning—a universal fear that existed not only within himself but exists within all of humankind.

There were also similarities in the prejudices each scientist had to overcome. For example, both explorers shattered our human narcissism—Darwin by linking us to animals, insects, and the like, and Freud by assigning much of our fate to unconscious forces beyond our conscious control (Sulloway, 1979). Both men also ultimately were brushing up against the fundamental and universal human dread of personal death and mortality. Darwin did so by challenging the belief that humans have immortality and a special place in God's eyes. He also dealt with death through his conjectures about extinction, but his approach was depersonalized because he applied the concept mainly to species. Freud, on the other hand, dealt with individual issues of personal demise, although he did so somewhat defensively. Thus, Freud never properly recognized that in fashioning psychoanalysis he was dealing with emotional dysfunctions that were deeply and unconsciously grounded in death anxiety. Nor did he understand that his and his patients' psychological and other defences were in part erected to deny personal mortality—an alignment that is challenged on some level by psychoanalytic forms of therapy.

Neither man dealt with an unmixed picture. While Darwin helped us to celebrate the wonders and beauties of nature, he also presented us with predators and their prey, the violent ways of nature, and the ruthless (non-feeling, automatic, biological) ways of natural selection. And while Freud opened the door to the psychological study of human invention, imagination, and creativity, he also brought us closer and closer to all that is repugnant and awful in our emotional makeup.

Nevertheless, a key difference remains: Darwin studied nature as it presented itself to him without reservation, while Freud tried to penetrate a specific aspect of nature—the unconscious mind—that is hidden from us all and that is, as we will see, by natural design, kept from our view for adaptive reasons. Darwin's work flowed with nature, while Freud's work strove to undo a naturally defensive configuration. There is nothing essential to the architecture of the emotion-processing mind that would preclude Darwinian insights—their province, whatever their emotional impact, is more cognitive than emotional. But Freud's quest was being carried out in defiance of the defensive and self-protective architecture of the emotion-processing mind, a universal and personal architecture that he could not readily overcome.

Psychoanalysis appears to be the only biological science that the human mind is designed to not undertake. It is a science that goes against the grain of the highly defensive, evolved design of the emotion-processing mind.

With this as our perspective, let us turn now to the task of developing a framework for our assessment of Freud's interaction with Darwinian theory.

CHAPTER THREE

A communicative framework for evaluating Freud

There have been many assessments of Freud's interaction with Darwinian theory, each from its own vantage point. Most psychoanalytically oriented writers (e.g. Badcock, 1994; Ritvo, 1990; Slavin & Kriegman, 1992; see also Sulloway, 1979) have characterized Freud as having adopted a strong adaptive and evolutionary viewpoint. They have stressed the spirit of harmony between Freud and Darwin and pointed to the many ideas that Freud either borrowed from Darwin or developed independently in ways that concurred with Darwin's thinking. These writers have also tended to excuse the flagrant flaws in Freud's evolutionary position—mainly his life-long adherence to Lamarckism (the disproved concept of the inheritance of acquired characteristics) and recapitulationism (the idea that ontogeny recapitulates phylogeny—the almost entirely refuted proposition that individual development repeats the development of earlier species; see Dennett, 1995).

For example, both Sulloway (1979) and Ritvo (1990) stressed Darwin's positive influence on Freud, and each catalogued the many ideas—for example, the nature of affects, the importance of sexuality, the concept of regression—that Freud had borrowed from Darwin or for which Darwin implicitly offered Freud his considerable support. A similar attitude was adopted by Slavin and Kriegman (1992), who offered an epic presentation of evolutionary psychoanalysis that was organized around relational behaviours and their seeming genetic and evolutionary sources. While they advocated recasting classical psychoanalytic thinking into adaptive and relational terms, they nevertheless attempted to show the roots of their evolutionary approach in Freud's writings and to stress again the allegiances of Freud to Darwinian theory.

Badcock (1994) specifically proposed a unification of the two theories—Freudian and Darwinian—in what he calls *psycho-Darwinism*. However, he based his joining together of these two fields of biological science almost entirely on Freud's early writings, in that he argued that the revisions in Freudian theory made in the last fifty years are more a regression than an advance. This doubtful position should warn us of the treacherous issues that must be negotiated in developing an interdisciplinary amalgam and in assessing Freud's interaction with Darwinian concepts.

On the other hand, the main critic of Freud's evolutionary thinking was Kitcher (1992), a cognitive scientist who used Freud to illustrate the dangers and pitfalls of interdisciplinary research—Freud fell prey to many of them. She showed how the errors that Freud made in respect to Darwinian theory and the intransigence of his thinking were detrimental to his theoretical constructs. She was also well aware that errors can be made on either side of an interdisciplinary equation—in this case, within evolutionary or psychoanalytic theory, or both.

These writings reflect a familiar polarity in the biological and evolutionary realms—that of similarity versus dissimilarity, continuity versus discontinuity. Darwin, for his purposes, overemphasized the continuity of species to the disservice of discontinuities, especially as they pertained to *Homo sapiens sapiens*. Similarly, the psychoanalytic writers who have examined Darwin's influence on Freud have stressed similarities at the expense of dissimilarities. They have thereby missed some significant problems in Freud's evolutionary position.

To clarify this point, we may note that viewed from the vantage-point of the communicative approach (see below), the picture of Freud that emerges is that of someone who did not grasp or use the essentials of Darwin's theory of evolution, and who did, indeed, make a series of fundamental errors in articulating his version of evolutionary theory. In addition, Freud did *not* adopt the strong adaptive viewpoint that is a cornerstone of Darwinian thinking. These failings led to serious flaws in his psychoanalytic theory and techniques, although one could argue that it was the other way around—namely, that some need in Freud for a compromised psychoanalytic theory and way of doing psychoanalysis led him to misunderstand and misapply evolutionary principles.

The communicative perspective

Let us turn now to the framework—the version of psychoanalytic theory—on which my own discussion has been and will be based. Termed *the communicative or adaptational-interactional approach or viewpoint*, it is a fundamentally adaptation-oriented position. The following are its main propositions (see Langs, 1982, 1992a, 1993a, 1994a, 1995c; Smith, 1991):

1. Human beings such as patients and therapists are adaptive organisms who are designed to cope with their environments (and secondarily, themselves) through both physical and mental means.
2. Adaptation is evoked by and is centred on *immediate external stimuli* termed *triggering or trigger events* or *environmental impingements*. Internal, within-organism events—one's own inner thoughts, feelings, and physical sensations—are, as a rule, secondary adaptive issues.
3. In the emotional realm, adaptation takes place in two forms and on two distinctive levels, with and without awareness interceding—*consciously* and *unconsciously*.
4. The basic approach to *listening and formulating* the material from patients in psychotherapy calls for organizing their free associations and other communications in terms of their directly stated *manifest meanings and their implications*, and

their latent, disguised, or *encoded meanings*. While manifest meanings deal with a wide range of *conscious* adaptive issues, within and outside of therapy, encoded meanings essentially are constituted as *unconscious* adaptive responses to the implications and meanings of the immediate stimuli or trigger events facing patients—the interventions of their therapists, including silences, comments, and behaviours, and especially their management of the ground rules and conditions of their therapies.

    a. This listening process culminates in a process known as *trigger-decoding*, through which the therapist *links* the patient's extracted narrative themes to their adaptation-evoking trigger events. This is, at present, the only known means of accessing *deep unconscious experience and adaptations*.

5. Successful adaptations require sensitive and reliable sensory-perceptual apparati and effective conscious processing of incoming information and meaning. Organisms who consistently misperceive the nature of their environments, or who are dominated by projections onto others rather than accurate perceptions, are unlikely to survive, nor would they be expected to have a strong degree of reproductive fitness. Emotionally, however, it appears that *conscious perception* is a compromised faculty, while *unconscious perception* is relatively unimpaired.

6. *Homo sapiens sapiens* has evolved with two distinctive methods for processing emotionally charged environmental impingements:

    a. via *affective and physical responses* whose *inherent meanings* are expressed in global and ill-defined fashion;

    b. via an emergent and unique *mental/communicative cognitive capacity—the emotion-processing mind*. This is the functional mental module that processes incoming emotionally charged messages and their meanings by means of a complex, two-tiered, parallel-processing, language-based adaptive system (see Bickerton, 1995).

7. The emotion-processing mind is configured as a two-system module for adapting to emotionally charged information and meaning (for details, see chapters 7 and 8 and Langs, 1995a).

a. The first system is focused on immediate and long-term issues of survival. It is called the *conscious system*, and it receives messages within awareness, engages in what is largely conscious adaptive processing efforts, and then responds directly and knowingly to existing impingements.

b. The second system is termed the *deep unconscious system*. It receives messages *subliminally, without awareness*, registers their meanings unconsciously, engages in adaptive processing entirely outside of awareness, and then expresses the results of this processing through *encoded (disguised) narratives*.

8. In psychotherapy, the primary adaptive tasks or triggers for psychotherapy patients are constituted by the interventions of their therapists. Similarly, the primary adaptive tasks for therapists are the moment-to-moment communications, behaviours, affects, symptoms, and clinical remissions of their patients. Stimuli and impingements that occur outside of therapy are secondary adaptive issues.

   a. The most powerful class of emotionally charged triggers for *unconscious* adaptive responses as revealed through their encoded narratives are almost always *frame-related*. For patients, these are interventions that are constituted as their therapist's management of the ground rules, setting, and other conditions of the treatment situation. Therapists' valid interventions or errors and the level of meaning at which a therapist works also serve moderately to organize *unconscious* adaptive communications from patients.

   b. Years of experience indicate that *encoded, unconsciously validating narratives and images from patients* speak for a *universal, ideal secured frame* for a psychotherapy experience that holds the patient safely—and the therapist. This frame includes a single setting, an agreed fee, and a fixed time and frequency of sessions; a one-to-one relationship without the intrusion of third parties; total privacy and confidentiality; and the relative anonymity of the therapist. Also involved are his or her use of trigger-decoded interventions (interpretations and frame-securing efforts that stem from the patient's encoded, unconscious perceptions of the therapist's prior frame-related and other efforts). All such work requires encoded validation for its confirmation (Langs, 1992a, 1993a).

c. *Secured frames* are deeply enhancing but are also the cause of *secured-frame anxieties*, which are linked to personal death anxiety, a fundamental human existential dread and issue. As a result, secured frames are feared consciously, although welcomed deeply unconsciously.

d. Departures from these ideal ground rules create *altered or deviant frames*, which are harmful to patients and therapists alike but generally favoured by them because of defensively motivated conscious-system preferences. Deviant frames are *pathologically* gratifying and, at times, temporarily self-protective; while they tend to reduce the secured-frame anxieties of both parties to therapy, they do so at great cost to all concerned.

The key precepts

The main precepts of the communicative approach to psychotherapy can be summarized as:

1. Patients and therapists are first and foremost adaptive organisms.
2. The immediate environment, broadly defined, is the primary source of emotionally charged trigger events for the adaptive responses of each party to therapy.
    a. Thus, the central adaptive issues for patients and therapists arise in the here-and-now therapeutic interaction and the vicissitudes of its ground rules and frame.
3. The adaptations made by patients and therapists are carried out through *both conscious and unconscious* means and processes.
4. Patients' narrative material is two-tiered in that its manifest contents reflect direct responses to conscious adaptive is-

sues, while its encoded contents are responsive to unconsciously perceived adaptive issues. As a rule, these two sets of reactions tend to be very different in a host of ways (see chapters 7 and 8).

The communicative approach and evolution

The communicative approach possesses several unique features that favour its use as a means of linking psychoanalysis with evolutionary theory, and as a basis for assessing Freud's evolutionary thinking. The following points are pertinent:

1. The communicative approach is essentially an adaptationally formulated theory and therefore naturally intersects with the adaptive foundation of evolutionary theory.

2. In establishing adaptation to environmental impingements as the primary activators of unconscious intrapsychic and interpersonal emotional-coping responses, the communicative approach facilitated the discovery of the basic organ of psychological adaptation—the emotion-processing mind. This fundamental processing and adaptive system lends itself readily as a natural unit of selection for studies of both adaptation and evolution. As such, this entity greatly facilitates the synthesis between psychoanalysis and evolutionary biology.

    a. As presently conceptualized, the model of an emotion-related cognitive-processing system is more specific, readily decomposed into operationally meaningful segments, and generally serviceable for the investigation of emotionally charged adaptations than the Freudian model of a tripartite mental structure consisting of an ego, superego, and id (Freud, 1923b). The communicative model addresses the basic adaptive task of processing information and meaning and is hierarchically more fundamental than any of the more complex structural entities defined by Freud. Indeed, each of these more complicated entities makes use of aspects of the functioning of the emotion-processing mind, whose faculties are basic to their operations.

    b. The fundamental axiom of psychoanalytic theory involves the *unconscious* domain. Unconscious communication, experience, and adaptive processing are empirically defined in communicative theory and are at the heart of its model of the mind. This version of psychoanalysis is, therefore, ready to meet with evolution as a theory that fully embraces the core concept of analytic thinking.

3. As an observation-driven theory, the communicative approach essentially is a selectionistic theory and therefore readily intersects with the selectionistic principles of the science of evolution.

    a. As an interactional and dialectical theory, the communicative approach also readily embraces the Darwinian principle that adaptation, learning, and acquisition of knowledge are *interactional*. Accepted too are the ideas that the environment is selectively experienced by the organism (Piaget, 1953, 1979), and that the organism, in turn, influences the environment through its behaviours. The organism and its environment are therefore a systemic entity with emergent systemic properties; there is a prevailing sense of an ever-present mutual interaction. These adaptive/evolutionary concepts are also part of the basic communicative theory.

4. Finally, the communicative approach is the only hierarchically structured theory of psychoanalysis, an attribute that also affords the approach a strong position in seeking a meeting ground with evolution.

Overall, then, the communicative approach shows a large number of isomorphisms, and considerable compatibility, with evolutionary theory. This seems to qualify the approach as a sound basis for assessing Freud's evolutionary position and for a later integration between psychoanalysis and Darwin's science of evolution.

CHAPTER FOUR

Freud's evolutionary thinking

I have set the stage for a fresh evaluation of Freud's interaction with Darwinian theory. Almost all of the essential propositions of evolutionary theory were articulated during Freud's lifetime (Kitcher, 1992; Ritvo, 1990; Sulloway, 1979). Darwin's overriding principles of competition between species for survival within given environments and descent with modification via natural selection were well established in the late nineteenth and early twentieth centuries. In 1893, Weissman showed that germ or sex cells, as distinguished from somatic cells, are solely responsible for transmitting hereditary information, dealing a death blow to the Lamarckian theory of the inheritance of acquired characteristics via somatic cells. Recapitulationism—Haeckel's thesis that ontogeny recapitulates phylogeny—fell to refutational evidence from experimental embryology (though it remains a crude approximation; see Dennett, 1995). And Mendel's work with the mechanisms of inheritance and the discovery of the gene were additional realizations that found extensive acceptance in the first two decades of the twentieth century.

As Kitcher (1992) has most clearly shown, Freud maintained a peculiar intransigence in his evolutionary perspectives. He selectively adopted a number of ideas that were in vogue during the years he trained as a scientist, Lamarckism and recapitulationism chief among them. But then, despite their refutation, he adhered to these concepts throughout his career as a psychoanalyst. In his writings he sustained *a weak, though persistent evolutionary and adaptive vantage point* whose specifics were ill-defined and often in error. And although he borrowed a number of key ideas about human emotional functioning and the psychoanalytic situation from Darwin (Sulloway, 1979), he never articulated an evolutionary history for his two psychological models of the mind (Freud, 1900a, 1923b), and he never included adaptation among his meta-psychological viewpoints.

The early Darwinian theory that Freud embraced was fraught with confusion. The basic principle of natural selection eluded Freud, who was inclined towards a Lamarckian, instructionistic way of thinking. Thus, Freud favoured the idea that the environment directs an organism's adaptations, rather than seeing the environment as selecting adaptations from resources already in place. Even today, when psychoanalysts are no longer Lamarckian in their evolutionary thinking, their clinical work unwittingly is infiltrated with Lamarckian instructionism (see chapters 15 and 16).

Freud's failure to appreciate the many facets of Darwinian selectionism is most unfortunate, but there seems to be a natural inclination for the conscious mind to be drawn to instructionism rather than selectionism. Indeed, instructionism has consistently been the first model for ideas about the adaptive functioning of various entities, such as the immune system and the brain. Darwin himself was not free of erroneous Lamarckian beliefs, including the principle of use and disuse—essentially, the inherited acquisition of used, learned behaviours and tendencies. Freud also made frequent use of this rejected concept, as seen most clearly in *Totem and Taboo* (Freud, 1912–13) and his work on a phylogenetic fantasy (Freud, 1915/1985).

Freud's use of evolutionary ideas

We turn now to the question of the extent to which Freud made use of and incorporated evolutionary precepts into psychoanalytic theory. Broadly speaking, a review of Freud's writings indicates that although he often alluded to evolution, he did so in a general manner, rather than considering the specifics of evolutionary theory. For example, Freud often acknowledged that psychic mechanisms have evolved over the millennia, but he never examined the implications of that proposition, nor did he make an effort to spell out the trajectory of that evolutionary history. In addition, as noted, the adaptive point of view was not part of Freud's meta-psychological theory—he confined himself to dynamics, personal genetics, economics, and psychic structure. And while he took from Darwin the idea of *sexual* selection and the importance of sexual reproduction in evolution, stressing and overstating the role of sexual conflicts in the aetiology of "neuroses," he made little use of *natural* selection and basic Darwinian principles in forging his main theories of psychopathology and of the nature of the analytic interaction and experience.

There are only twenty references to Darwin and his biological work in all of Freud's writings (Ritvo, 1990). This is a surprisingly small number of allusions, given the fundamental place of Darwinian theory in biology and psychology and the size of Freud's corpus. This sparseness and the lack of any broad effort to unite psychoanalysis and evolution is in keeping with the realization that most of the basic propositions of psychoanalytic theory were not constrained by—nor developed to any significant degree in light of—the Darwinian theory of evolution, and they were fashioned with little attention to the shaping role of reality and the environment in the aetiology of "neuroses".

Clinically, Freud paid little attention to the realities of analysts' interventions and their impact on their patients. Instead, he proposed the concept of transference—the thesis that patients distort their views of their analysts based almost entirely on pathogenic early childhood experiences and projections of their own inner fantasies and memories onto the person of the analyst whose essential role is that of an impassive recipient. This line of thought runs counter to the stress in evolutionary theory on the role of reality impingements in adaptation and on the importance of sound perceptiveness in adapting to environmental impingements.

By way of contrast, the communicative approach, which is primarily an adaptive theory, has shown that for patients, the interventions of their therapists constitute their primary adaptive tasks, especially with respect to *unconscious* adaptive responses. As for patients' distortions, adaptive listening has revealed that while manifest, consciously communicated distortions of the interventions of therapists do materialize in psychotherapy, their primary source is not a displacement from the past but, instead, derives from the staunchly defensive posture of the conscious mind in the face of disturbing unconscious perceptions of a therapist's errant efforts. Distortion typically is motivated by wishes to protect the therapist and the patient from disturbing conscious realizations of valid unconscious experiences of the therapist's errors—links to past figures, which do exist, are secondary.

The effects of the Freudian heritage are with us today even as psychoanalysts shift from his so-called one-person, projective psychology to a more integrative two-person psychology (see Gill, 1994; Slavin & Kriegman, 1992). Much of this revised theory remains concentrated on the patient's inner mental life and how patients' memories and fantasies influence their experiences of their therapists' interventions, which are all too vaguely defined. Interaction is acknowledged, but there is a lack of the kind of Darwinian specificity that calls for a detailed examination of the nature of specific external (interventional) impingements and their definitive universal meanings. Each patient will personally and unconsciously respond to these shared meanings by selecting those that are pertinent to his or her emotional state and life. In addition, the patient will react to an intervention in terms of idiosyncratic meanings as well—both selected universals and personal sensitivities play a role.

This is especially true of the *unconsciously* conveyed meanings of a therapist's efforts—especially his or her management of the ground rules of therapy. The entire level of environmentally evoked, *unconscious, and encoded communication* within psychotherapy is neglected by these therapists, leaving them with a one-dimensional, flatland picture of human communication and the therapeutic interaction and experience (Abbott, 1884).

Freud's mis-readings of evolutionary theory are illustrated in his 1915 paper, "Instincts and their Vicissitudes" (Freud, 1915c). In the part of the paper I cite, he was arguing that the nervous system functions to get rid of and maximally reduce the intensity of the stimuli that reach it. (This is a reference to the so-called *principle of constancy*, an untenable and subsequently abandoned proposed biological principle related to Freud's early ideas regarding mastery.)

In the paper, Freud goes on to state that organisms are capable of withdrawing from external stimuli, and that, as a result, the ability to escape from outer threat became a hereditary disposition (a clear Lamarckian proposition). On the other hand, he continues, instinctual stimuli, in that they arise from within the organism, cannot be dealt with by this mechanism. (This too is an untenable proposition in that flight-related psychological mechanisms, such as denial and repression, are in ever-present use as defences against what Freud termed instinctual stimuli—the latter a somewhat confused term that Freud used to refer to inner drives and wishes.) Instinctual drives, therefore, by Freud's reasoning, make more demands on the nervous system than do environmental impingements. (This is another questionable assumption, one that violates Darwinian principles.)

Freud then sums matters up by stating:

> We may therefore well conclude that instincts and not external stimuli are the true motive forces behind the advances that have led the nervous system, with its unlimited capacities, to its present high level of development. There is naturally nothing to prevent our supposing that the instincts themselves are, at least in part, the precipitates of the effects of external stimulation, which in the course of phylogenesis have brought about modifications in the living substance. [Freud, 1915c, p. 120]

Freud's line of thought is laced with assumptions and arguments that range from questionable to clearly erroneous—not the least of which is the common mistake of mixing levels that should be distinct—that is, confusing mind and brain. He reasons that instincts rather than the environment are the *primary driving force and selection factor* in evolutionary change. He also assumes that instinctual pressures are inescapable as compared to environmental pressures, and he argues for phylogenetic or Lamarckian causes of evolution. Although there are some indications that inner needs may exert a small degree of selection pressure, by and large all of these ideas are untenable. They appear to reflect Freud's failure to grasp the essentials of Darwinian theory and principles of evolution. These deficiencies have had notable detrimental consequences for psychoanalytic thinking, including its entry into the adaptive and evolutionary realms—consequences we must now strive to correct.

Some specific borrowings

While Freud evidently did not comprehend the fundamentals of evolutionary theory, he did appropriate many aspects of Darwin's writings (Badcock, 1994; Ritvo, 1990; Sulloway, 1979). For example, Freud made use of aspects of Darwin's elaborate theory of affects (Darwin, 1872; Nesse, 1990a), which are seen today as non-verbal, adaptive means of knowing, responding to, and affecting environmental impingements (Plotkin, 1994). From Darwin, Freud also derived the view of emotional dysfunctions as meaningful disorders that are forms of psychopathology or maladaptation. And the representation of affects through associated images is a Darwinian idea that supported Freud's thinking about the symbolic representation of affective states and, more broadly, the symbolic meanings of emotional symptoms. Darwin's concept of an overflow of excitations in emotionally charged situations contributed to Freud's economic meta-psychological viewpoint, and the notion of antithetical emotional states and meanings led Freud to ideas about mental defences such as reaction formation. The Darwinian concept of survival of the fittest and of competition for resources was a factor in Freud's strong investment in conflict theory. And Darwin's idea that evolution is a historical force and his thinking about developmental processes helped shape Freud's thinking on the importance of personal genetics—individual history—in the aetiology of emotional disorders.

Freud made little use of Darwin's key principle of *natural* selection, thereby neglecting economic and survival issues. He did, however, turn to the lesser principle of *sexual* selection. This principle played a notable role in the development of Freud's ideas about libido, the sexual instincts, and psychosexual development and contributed to his overemphasis on sexual instinctual drives in individual development and, especially, on the role of sexual conflicts in the development of neurotic symptoms. However, Freud did not distinguish sexuality as an individual intrapsychic and interpersonal adaptive issue from sexuality as a factor in the process of evolution.

Freud recognized Darwin's ideas about the relevance of both developmental and environmental factors in individual maturation. He transformed this idea into a complementary series in which the developmental-instinctual factors were afforded overriding emphasis over those that were environmental. Darwinian thinking also helped shape Freud's proposals regarding the proximal and distal—immediate and long-range—causes of neurosis. However, Darwin stressed distal causes, the shaping and selection powers of the environment, and the importance of inter-species and intra-species conflict, while Freud was preoccupied with proximal causes and immediate intrapsychic issues to the relative exclusion of all else—an emphasis that latter-day evolutionary psychoanalysts have been attempting to set right (Slavin & Kriegman, 1992).

While Freud missed the spirit of the Darwinian theory of natural selection, he did, as noted, embrace the Lamarckian theory of use and disuse. He also adopted Lamarckian evolutionary theory to buttress his psychoanalytic theorizing (see above and Freud, 1912-13, 1915/1985). In at least two instances, he imposed his theoretical ideas onto questionable anthropological data, trying to generate a sense of seeming validation that has not stood the test of time.

Freud's main argument combined Lamarckism and recapitulationism. He claimed that the current intrapsychic conflicts seen in neurotic individuals reflect and stem in part from actions carried out by so-called primitive peoples (Freud, 1912-13). Using Darwin's concept of primal hordes, he argued that in ancient eras, conflicts were enacted in reality and then passed down from generation to generation via Lamarckian principles of use and disuse, and through the inheritance of acquired characteristics. In time, these enacted wishes and conflicts were internalized and passed on through later generations as fantasies rather than realities.

Freud proposed, for example, that the modern-day male child's oedipal wish to murder his father and sleep with his mother, and the resultant guilt caused by the fantasy, had all been enacted by so-called primitive people in times past. On this basis, he claimed validation of the psychoanalytic theory of the Oedipus complex—an unfounded conclusion based on uncertain premises.

In all, then, even though Freud was aware of evolutionary principles in some broad fashion and advocated teaching evolution in psychoanalytic institutes (Freud, 1927), he did not have a basic commitment to sound evolutionary principles or to the adaptive viewpoint with respect to immediate efforts at coping. His overall psychoanalytic thinking is considerably clarified through these realizations.

Some latter-day efforts

Freud's distance from evolutionary principles has left us with a relative absence of adaptive thinking in contemporary psychoanalytic theory, a void that a small group of psychoanalytic evolutionary writers have recently been attempting to fill (see Langs, 1995b, for details). Using various versions of classical psychoanalytic theory, these contributions have concentrated on the development of adaptationist programmes for such entities as the Oedipus complex (Badcock, 1990a, 1990b), psychic defences (Lloyd, 1990; Nesse, 1990b; Nesse & Lloyd, 1992), a variety of relational issues with a focus on the problem of altruism (Badcock, 1986; Slavin & Kriegman, 1992), and a host of other phenomena (Nesse & Williams, 1994).

These efforts have relied heavily on non-human biological data and are deeply committed to the selfish-gene theory of

Dawkins (1976b, 1987) and others—a reductionistic theory that sees the replication of genes as the overriding driving force in evolution (Eldredge, 1995; Wesson, 1994). They have paid little attention to the emergent aspects of human emotional functioning, including language and culture (Bickerton, 1990, 1995; Liberman, 1991). Nevertheless, these evolutionary psychoanalysts have brought much-needed fresh perspectives to the field and offer new ways of thinking about a variety of psychodynamic and relational issues.

These early attempts to develop adaptationist programmes reflect some of the as yet unsolved problems that accrue to these much-needed efforts. For example, several of these writers (Nesse, 1990b; Nesse & Lloyd, 1992; Slavin & Kriegman, 1992) have attempted to trace the development of the psychic mechanism of repression to the selection advantages of deception and the cheating of conspecifics (others of the same species). Such cheating is said to require that the cheaters not be aware—repress—the fact that they are cheating lest they give themselves away and lose the reproductive advantage they are seeking to gain (Alexander, 1979; Nesse, 1990b; Trivers, 1976, 1985). However, this formulation does not appreciate the fundamental properties of repression as a basic, far-ranging psychological and interactional mechanism that has a variety of motives and several different forms. Nor does it account for the fundamental anxieties and unconscious perceptions to which humans automatically respond with unconsciously invoked repression to protect both themselves and others from dysfunction and interpersonal disturbance (see chapter 8 and Langs, 1995b).

Other areas that have preoccupied these writers are those of parent–offspring conflicts, self realization, and human altruism—with extensions of these ideas into the therapeutic interaction (Slavin & Kriegman, 1992; Trivers, 1971, 1974). Here, too, data from non-human species play too large a role to the neglect of abundant human data. As a result, the nature of conscious and especially unconscious psychic functioning is neglected. The emergent aspects of human emotional adaptations (Wesson, 1994)—such as the role of conscious and unconscious guilt in human altruism (Liberman, 1991; Langs, 1995b)—are the kinds of factors that deserve fuller consideration.

Despite the limitations of these initial efforts, evolutionary biologists and psychoanalysts are bringing evolutionary science into psychoanalysis and establishing a place for evolutionary concepts and insights in psychoanalytic theory and practice. The work presented in this volume offers an alternative view of the place of evolution in psychoanalytic thinking and, specifically, adds the dimension of human psychic functioning—the structure and adaptations of the emotion-processing mind—to these venturesome evolutionary probes. In time, evolutionary psychoanalysis should vitally enhance psychoanalytic theory.

Psychoanalysis without biology

Several major deficiencies in the fundamental theory of psychoanalysis have persisted because most analytic theorists have not incorporated the principles and constraints of evolutionary theory and biology into their thinking. The following seem most pertinent:

1. Despite its biological trappings and insistent allusions to instincts and other physiological phenomena (Sulloway, 1979), psychoanalysis did not develop as a branch of biology or as a biological science. It was therefore unnecessarily disconnected from the basic science of which it is, indeed, a subscience.

2. Because it eschewed biology in general and evolutionary theory in particular, Freud's psychoanalytic theory overemphasized the intrapsychic and instinctual-drive–related issues in human emotional life to the neglect of the role played by relational interactions and reality impingements. The shift in analytic theory towards relational issues and interpersonal conflicts has not as yet fully set matters straight—mainly because the personal biases of the perceiver are over-stressed as a factor in subjective experience, to the neglect of the factual meanings of environmental impingements, and because deep unconscious communication and adaptation have been ignored.

a. The non-biological approach allowed psychoanalysis to founder as a hermeneutic theory of inner meaning and fantasy. Psychoanalysis has been relatively detached from the biological core of mental functioning and from the interplay between the organism and realistic, environmental adaptive issues. Psychological meaning was seen as determined personally rather than through a balanced interaction between the actual, universal meanings of environmental impingements and the individual's selective conscious and unconscious perceptions and interpretation of these events—inner experiences that are constrained by the nature of the external event.

b. The psychoanalytic theories of neurosogenesis and of the nature of the therapeutic process similarly do not include the well-supported conceptualization that all knowledge acquisition, learning, and adaptation is interactional and dialectical, matters of assimilation and accommodation (Piaget, 1953, 1979). For example, the actual properties of the therapeutic environment and the specific meanings of particular interventions interact consciously and unconsciously with the patient's inner mental world to produce the flow of his or her therapeutic experiences.

c. A likely source of the neglect of environmental impingements comes from the psychoanalytic situation in which Freud and his followers defensively have not appreciated the multiple levels and full range of meanings conveyed in their interventions to their patients—many of them unconsciously transmitted and highly traumatic and harmful for the patient. These detrimental features of accepted psychoanalytic technique are revealed fully only through a strong adaptive approach to formulating the material from patients. The relevant insights materialize mainly through the use of trigger-decoding—deciphering the patient's imagery in light of the currently active interventions to which he or she is adapting. These adaptation-oriented formulations and their encoded validation reveal a remarkable range of *unconscious* meanings that accrue to the communications from therapists and the extent of patients' *unconscious perceptiveness and responsiveness* in the treatment situation.

3. Without a fundamental biological orientation and adaptive commitment, psychoanalysis has claimed exemptive status from science and has maintained itself with few scientific standards or principles—a truly disastrous position. This has opened the door to concepts and treatment practices that not only are without scientific support and validation, but are error-prone and harmful to patients.
4. Without science there are no unsolved puzzles or anomalies in psychoanalysis—intractable problems that frustrate analysts and confront them with the definitive limitations of psychoanalytic theory and explanation, and the damaging aspects of analytic practice. As a result, there is little motivation, especially among clinicians, for research and change.
5. The absence of a broad biological perspective also supports the psychoanalytic stress on the individuality of patients and therapists, and their particular therapeutic interactions. All but lost is a consideration of and search for universal factors (Brown, 1991; Tooby & Cosmides, 1990a, 1990b)—regularities, laws, and other consistencies across individuals, patients, and therapists and the systems they create together. Differences, distinctions, and divisions almost entirely dominate unity, communality, and synthesis—individuality obscures universality.
6. Without biology and evolution, there is little incentive for psychoanalysts to search for the fundamental adaptive and learning structures of the human mind and of human relatedness. Ill-defined, unmanageable high-level concepts of two kinds dominate the field—first, those that are clinical in nature, such as transference, countertransference, and resistance, and, second, those that refer to psychic structures like the ego, superego, and id.
    a. Psychoanalysis seems to be caught up in an early evolutionary phase of thinking similar to those passed through by other sciences such as biology, chemistry, and physics. The phase is marked by the use of ill-defined global terms like protoplasm and atmosphere that are used to describe highly complex, specifically configured entities in vague ways that do not foster scientific thought or investigation. The language appears to be scientific, but metaphor and pseudo-explanation prevail.

7. By not keeping abreast with developments in biology, psychoanalysts are highly mechanistic in their viewpoints, with a stress on fixed entities and mechanisms rather than on both structure and process, as well as change, self-organization, reciprocity, and dialectics. Criticisms of Freud's mechanics have not rectified this situation but, instead, seem to have led to new versions of the mechanistic viewpoint.

8. In the relative absence of biology and evolution, psychoanalysis has failed to gain a much-needed place in the family of sciences and has not established a clear identity of its own. The foundation theorem of psychoanalysis states that *unconscious* functioning, processing, communication, and adaptation are critical components of human adaptations in the emotional domain. This proposition played a significant role as a distinguishing feature of the systems of the mind in Freud's topographic definition of the mental apparatus (Freud, 1900a). However, in time, unprotected by scientific research and a biological–evolutionary orientation, the conception of the unconscious domain has become relatively inconsequential and lacking in precise definition—much to the loss of the field.

The term *unconscious* as used today alludes not only to mental contents, processes, or structures, but it has been generalized to include inclinations to act or relate in certain ways, personal and biological genetic make-up, automatic and unthinking behaviours, bodily functions, the operations of the brain, the workings of natural selection and other biological regularities, biological laws, and anything purported by one person to be absent from another person's awareness. No longer the fundamental and distinguishing theorem of psychoanalysis, the concept of an unconscious realm—once the hallmark of psychoanalysis—is all but lost. A fresh approach to this concept that incorporates adaptive and evolutionary considerations may well provide the field with the distictive identity it now seems to lack.

The many negative consequences of the estrangement of psychoanalysis from biology and Darwinian evolution speaks strongly for the need for a major rapprochement that will bring psychoanalysis into its rightful and necessary place as a biological science. Because science can serve as a relatively reliable guide and constraint on therapeutic craftsmanship, a science of psychoanalysis appears to promise the field a much-needed sound and serious set of propositions.

To take us further in these directions, let us turn now to creating a hierarchical theory of psychoanalysis to render it as complete a theory of the emotional domain as presently possible.

PART II

SOME
BIOLOGICAL PRINCIPLES
FOR PSYCHOANALYSIS

CHAPTER FIVE

Hierarchies in psychoanalysis

Three more introductory considerations are needed to provide a solid base for the evolutionary probes that are to follow. The first is a presentation of the hierarchical structure of psychoanalysis, which will define a psychoanalytic theory that has substance and depth. The second is a hierarchical theory of evolution that has sufficient depth for its alliance with the psychoanalytic domain. And, finally, we will need a clear picture of our unit of selection (Lewontin, 1970)—the emotion-processing mind.

The relevance of hierarchies

The commanding theories in biology must be hierarchical in design—layered and structured rather than unidimensional—because living organisms, as well as their systems and processes, are, with few if any exceptions, structured in multilayered fashion (Dawkins, 1976a; Eldredge & Grene, 1992; Eldredge & Salthe, 1984; Gould, 1982; Grene, 1987; Plotkin, 1994; Salthe,

1985). Thus, a biological theory that is all-encompassing and generalizable must account for and use explanations that embrace each of the many levels of functioning and causality that apply to its entities. This means that both evolutionary and psychoanalytic theories, if they are to be the overarching and most powerful theories of their respective and conjoint realms, must be multilevelled in their areas of observing and theorizing.

Efforts to define the hierarchies of both evolutionary and psychoanalytic theories are in their infancy. The precise hierarchical structure of the former is currently a matter of controversy (Eldredge, 1995; Plotkin, 1994), while that of the latter has barely been examined (Wilson & Gedo, 1993).

For both evolution and psychoanalysis, there is a multitude of meaningful hierarchies to which they belong. There are, for example, three overarching hierarchies of relevance to both fields.

1. The hierarchy of the *basic sciences*—and the place of the science of *biology* therein.
2. The hierarchy of the *biological sciences*, which includes *evolution* and *psychoanalysis* within its confines and locates them with respect to each other.
3. The hierarchy of the levels of organization, observation and meaning that exist respectively within the biological subsciences of evolution and psychoanalysis, and any additional hierarchies that emerge when the two fields are joined together to form a comprehensive unified theory.

    For psychoanalysis, I will refer to its most critical internal organization as a hierarchy of basic subtheories that combine to define the field in the most comprehensive manner possible. However crucial, this is only one of the many hierarchies that apply to the field.

The basic task

The first task confronting any science is to develop its own hierarchical structures as fully as possible. For psychoanalysis, an initial and limited effort in this direction can be found in a recent volume on this subject (Wilson & Gedo, 1993). The hierarchies described by the contributors to this volume essentially are of two kinds. The first pertains to issues of psychodynamics, while the second involves personal genetic development. The latter deals with the organization of cognitive and psychosocial-psychosexual phases as they shape childhood and adult intrapsychic structures and functions.

These studies involve investigations of the complex internal structure of but two of the ten basic subtheories that have been identified for psychoanalysis to date—the psychodynamic and personal genetics levels. Another kind of hierarchy explored in that volume deals with the relationship between psychoanalysis and other biological sciences, especially cognitive science and neuroscience—a hierarchy that pertains to interdisciplinary research, including Freud's thinking in this area (see especially Grossman, 1993).

The basic problem

A fundamental reason why psychoanalysis has been unable to assert itself as the commanding theory of the emotional domain (Gill, 1994) is the absence of a multidimensional psychoanalytic theory. At present, mainstream analysis essentially is organized around two levels of observation and thought—psychodynamics, which includes self concepts and dynamically defined interpersonal and relational transactions, and personal genetics, which deals with developmental issues. As shown below, a more complete theory of psychoanalysis embraces at least eight other levels of conceptualizing, each adding distinctive sets of insights that are vital to the basic corpus.

The nature of hierarchies

Hierarchies capture and help to organize the complexities of biological nature. They involve the ordering of entities according to such criteria as scale, influence, dominance, power, functions, energy level, size, importance, and the like. There are two basic types of hierarchies (Plotkin, 1994; Salthe, 1985):

1. *Structural or nested hierarchies* are characterized by *containment* so that one entity is nested or contained within another entity, or one entity within a system builds its functioning on the basis of another, more fundamental entity. This dependency and nesting can repeat itself on as many deeper or more basic levels as actually exist in the system, until some seemingly irreducible, fundamental level is reached.

   An example of this type of hierarchy is an inanimate object like a chair, which is most fundamentally made of quarks and other basic particles or forces, which are nested within subatomic particles like electrons, protons, and the such, which are nested within atoms and molecules, which are nested within wood slats, which are nested within the chair itself. Another example is the human body, which at the deepest level is also made up of subatomic particles and, from there, molecules, cell parts, cells, tissues, organ systems, and the body as a whole.

   Hierarchical thinking organizes living systems but is fraught with issues of its own, such as defining the key hierarchies, their boundaries (nature's joints, as they are called), and their own internal ordering.

   In evolutionary science, there is a nested hierarchy that begins with the basic atomic structure of RNA and DNA and moves up from there to genes and chromosomes, individual development, phenotypes, cognitive functions, species, populations, socio-cultural transactions, and ecosystems. In psychoanalysis, the communicative approach has shown that psychodynamics and personal genetics are nested within a hierarchy that includes science, systemic features, relatedness, communication, and ground rules and frames (see below).

   A crucial feature of nested hierarchies is that, in general, the more fundamental levels exert the most powerful effects and constrain the operations of the levels lying above them. That is, higher levels on a hierarchical scale are not only dependent on lower levels, but cannot violate any principle that pertains to the more fundamental levels in the hierarchy.

2. *Control hierarchies* are the second fundamental type of hier-

archy. They are ordered according to authority and power, and they involve the dynamic passing of information between levels. Military organizations and universities are ordered as control hierarchies. This type of hierarchy can be found within the systems of a given individual as well as within a wide range of organizations, and it is a common feature of the relationship between competing individuals or organizations as well.

In psychotherapy, control hierarchies are made up of the many forms of treatment, of therapists and their patients (in that order), and of the organizations of training institutes. Two other control hierarchies of note are, first, the ordering in which the emotion-processing mind is seen as basic to the structures and processes of the mind—id, superego, and ego, including relating and communicating, and, second, the hierarchical structure of the emotion-processing mind itself, including the relationship between its conscious and deep unconscious systems. In addition to their control aspects, both of these psychoanalytic systems have structural or nested hierarchical features as well. Intermixtures of the two types of hierarchy are common, and the delineation of the features of each kind of hierarchy reveals distinctive information about the nature and adaptive functioning of the entity in question.

Scientific hierarchies for psychoanalysis

The communicative approach to psychoanalysis has been compelled by its confrontations with inexplicable clinical observations—so-called unsolved puzzles or anomalies (Langs, 1992a, 1992b, 1992c, 1993a, 1993b, 1995a, 1995b; Kuhn, 1962)—to develop a multi-hierarchical theory for the field. At the heart of this theory is the aforementioned hierarchy of basic subtheories that combine to articulate a comprehensive explanatory theory. To date, the approach has identified at least ten relatively independent yet related hierarchical systems pertinent to core psychoanalytic theory—each with its own mode of observation, structures and processes, rules of operation, and theoretical constructs.

Before ordering the ten basic subtheories of psychoanalysis, the following additional hierarchies deserve mention:

1. As a biological science of the emotional domain, psychoanalysis is hierarchical in at least two ways:
   a. Within its own scientific subtheory, there are at least three distinctive, nested forms of psychoanalytic science (see also Langs & Badalamenti, 1992b), which are, beginning with the most basic:
      i. the investigation of emotional cognition;
      ii. the study of emotionally relevant individual development;
      iii. the study of affects and their vicissitudes (Nesse, 1990a; Tooby & Cosmides, 1990b).
   b. Within the hierarchy of sciences as a whole (Kitcher, 1992), the nested sequence appears to be:
      i. physics and astronomy;
      ii. chemistry;
      iii. biology, which includes
         (1) evolutionary theory
         (2) physiology and neuroscience
         (3) psychology, which has its own hierarchy of subsciences, including psychoanalysis.
   c. Thus, psychoanalysis, as the science of emotional expression and cognition, is nested within the biological sciences, and biology is nested within physics and chemistry. Evolutionary psychoanalysis is nested within evolutionary biology, which in turn is nested within biology proper. Furthermore, in addition to their nested qualities, these hierarchies have control features as well, with physics having claim to its place at the most basic level of the control chart and psychoanalysis at the least basic.
   d. The existence of these hierarchies indicates that psychoanalysis is, can be, and must be a biological science and not merely a matter of hermeneutics and meanings devoid of adaptive implications—a position that is both anti-scientific and solipsistic.

e. Evolutionary psychoanalysis has considerable power within this hierarchical structure in that it is a branch of the most fundamental subscience of biology—evolutionary biology, the study of adaptations and their histories. This affords considerable importance to evolutionary principles as they apply to our unit of study—the emotion-processing mind and its functions.

An ordering of the basic subtheories of psychoanalysis

The fundamental theory of psychoanalysis contains a series of subtheories that can be organized as both a control and a nested hierarchy. I list here the ten *sub-theories* that have been identified by the communicative approach to this point in time, doing so in a sequence that begins with what appears to be *the most powerful and basic sub-theory* and moves up the scale from there.

1. *A formal science of psychoanalysis*, which includes all types of quantitative, mathematically grounded scientific efforts to measure psychoanalytic data and derive the regularities, laws, and rules of the mind—and of emotionally charged human communication and interactions (Langs, 1992c; Langs & Badalamenti, 1994a. 1994b, in press).
2. *The science based on stochastic and statistical mathematical models*, whose methods can define deep regularities and patterns of adaptation, but cannot establish deep laws (Shulman, 1990; see also Langs & Badalamenti, 1992b).
3. *Evolutionary psychoanalysis*, including the study of adaptation and the creation of historical, explanatory adaptationist programmes (Badcock, 1986, 1990a, 1990b, 1994; Glantz & Pearce, 1989; Lloyd, 1990; Nesse, 1990a, 1990b; Nesse & Lloyd, 1992; Nesse & Williams, 1994; Slavin & Kriegman, 1992).
4. *The systems theory* of psychoanalysis (Langs, 1992c).
5. *The science of psychoanatomy or model making*, the investi-

gation and definition of the architecture and processing capabilities of the emotion-processing mind (Langs, 1995a).

6. *The science of emotionally pertinent contexts, frames, and settings*, including the ground rules of the analytic and therapeutic situations (Langs, 1982, 1988, 1992a, 1993a, 1994a).

7. *The study of interactions, relating, and relationships* (Langs, 1992a, 1993a, 1994a; Slavin & Kriegman, 1992).

8. *The science of emotionally related personal development or personal genetics* (Wilson & Gedo, 1993).

9. *The exploration of psychodynamics*, including issues of conflict, self, and identity, and the study of gross psychic structures and their functions (Gill, 1994; Lloyd, 1990; Nesse, 1990b; Nesse & Lloyd, 1992; Slavin & Kriegman, 1992).

10. The investigation of human emotions and affects (Badcock, 1994; Nesse, 1990a; Tooby & Cosmides, 1990b).

Each hierarchical level has its own purview, areas of observation and means of observing, rules and regularities, postulates, theoretical precepts, and means and types of influence. Each level therefore generates and deals with particular data and develops distinctive theoretical constructs that contribute to the overall psychoanalytic picture and theory. Indeed, none of these subsciences can be entirely subsumed under any of its confreres, although each level is constrained by the levels that are more basic to it. In addition, each level informs the others and is informed by them in return. Psychoanalytic theory is incomplete without an accounting from each member of its hierarchy of subsciences.

Some additional hierarchies related to psychoanalysis

The delineation of the above nested and control hierarchy enables us to define psychoanalysis in comprehensive and diverse terms. There are, however, several additional hierarchies related to psychoanalysis that need to be mentioned here. The first of these hierarchies includes the following, listed again in order of saliency and power:

1. *The formal science of psychoanalysis*, which has a nested hierarchy of its own.
2. *Clinical psychoanalysis*, with its theory of the therapeutic process and interaction, including the process of cure.
3. *Developmental psychoanalysis*, with its theory of human emotional development.
4. *The psychoanalysis of affects and emotions.*

There are, however, still more hierarchies of import to our pursuits. The following are of note:

1. There is a bi-levelled hierarchy composed of deep processes and structures versus emotionally charged surface behaviours (Kitcher, 1985; Slavin & Kriegman, 1992; Tooby & Cosmides, 1990b). Processes and structures are in the more powerful position compared to the behaviours for which they are responsible.
    a. There is also a thing–entity–structural level and a process–function–operation level, which form still another notable hierarchy related to human emotional adaptations.
2. There also is a hierarchy of *factors in and causes of emotionally related behaviours*—adaptive and maladaptive. In terms of a control hierarchy, with the most powerful elements listed first, these causes may be classified as:
    a. Proximal or immediate causes.
        i. direct or activating causes—immediate stimuli, environmental impingements, or triggers, such as the actions of others and natural events, including those that are salutary or traumatic;
        ii. contextual causes—settings, the nature of background relationships, the influence of social and cultural factors, and the role of laws, rules, and ways of framing behaviours and communications;
        iii. an individual's current adaptive resources, which are based on his or her genetic make-up and experienced developmental events, including both the universal and individual characteristics of the emotion-process-

ing mind (this is the inner causal factor that interacts with outer, environmental factors).
   b. Distal or past causes.
      i. historic or evolutionary events, which include the nature of long-past environmental conditions and forces, the so-called *fitness landscape or environment of evolutionary adaptedness*, as well as the history of the resultant adaptations;
      ii. the nature of an individual's genetic endowment;
      iii. an individual's personal developmental history, from *in utero* to his or her unfolding personal life cycle;
      iv. an individual's particular prior life experiences, especially those from the earliest years of his or her life;
      v. the evolutionary history of *Homo sapiens sapiens*, the species of which we all are members.
   This particular hierarchy sets the stage for a detailed investigation of the nature and sources of mental health and mental dysfunctions. Within each of these hierarchical levels, there are additional nested and control hierarchies, a reminder again of the complexity of the human mind and its emotional adaptations and maladaptations.
3. There is another hierarchy of elements that defines the therapeutic interaction, which includes, in order of power:
   a. The setting and ground rules.
   b. The history of the relationship and interaction between patient and therapist, including its origins.
   c. Recent events, including the communicative material from and behaviours of the patient, and the interventions and behaviours of the therapist.
   d. The immediate communications and behaviours of both patient and therapist in a given session and in relation to any recent contact outside of or between sessions, if it should occur.
4. There is a hierarchy of professional organizations and relationships that exists for each psychotherapist and psychoanalyst.
5. Another hierarchy involves the ordering of the more than

three hundred forms of psychotherapy that exist today. Among the many possible criteria that could be used for organizing this hierarchy, perhaps the most critical dimension for the present discussion is *the extent to which a given form of therapy gains access to deep unconscious experience and communications*. The resultant nesting is as follows, with the greatest access listed first:

   a. *Empowered psychotherapy* (Langs, 1993a), a form of communicatively based treatment in which narrative material is maximized to allow for the development of an especially large pool of themes that embody the encoded, unconscious perceptions of therapist-related events. This form of therapy also calls for an intense search for adaptation-evoking triggers. Interpretations are made by decoding the emergent encoded images and disguised themes and linking them to their immediate adaptation-evoking triggers—a process that results in the conscious realization of deep unconscious experience and meaning. Because of the intensity and thoroughness with which these pursuits are carried out, this mode of treatment provides the greatest degree of access to deep unconscious experience available today.
   b. *Communicative psychotherapy* (see Langs, 1982, 1988, 1992a), which is a less intense version of empowered psychotherapy.
   c. Forms of *psychoanalytically oriented psychotherapy*, which, however, seldom access deep unconscious meaning because the work is carried out in terms of manifest contents and their implications, postulated intrapsychic fantasies and memories, and easily decoded symbolic expressions that relate to the *superficial unconscious subsystem of the conscious system* (see chapters 7 and 8). A sense of both *unconscious adaptation* and the use of *trigger-decoding* are missing from these approaches.
   d. *All other forms of therapy* that make no effort to access unconscious meanings.

6. There is a hierarchy that applies to the fundamentals of human emotional adaptation that involves *conscious versus unconscious* communication, processing, experience, and

meaning. The key issue here is the hierarchical relationship between the conscious and unconscious realms and the conscious and deep unconscious systems. Such qualities as their respective power and influence organize this hierarchy, which, in turn, is critical to psychoanalytic understanding and to treatment strategies in psychotherapy.

    a. In some ways, the conscious realm is the more powerful (e.g. with regard to the choice of direct responses to stimuli), while in other respects the deep unconscious realm is more compelling (e.g. with regard to emotional motivational power, the accurate perception of emotionally related realities, and the wisdom of adaptive responses).

7. Finally, an important hierarchical relationship exists between the two basic of modes of human experience and interacting with the world—relating and communicating. Current evidence suggests that communication is basic to relating, and that together they constitute the essential features of human emotional adaptation.

While the list is certainly not exhaustive, it can be seen that the concept of a hierarchical structure for the vast domain of emotional experience and adaptation pertinent to psychoanalysis establishes a foundation from which we can, first, enhance psychoanalytic thinking in general and, second, develop fully the subscience of evolutionary psychoanalysis as a basic and vital subscience of the field of psychoanalysis. In order to develop a working model of this subscience, a presentation of the hierarchical levels and basic principles of the Darwinian theory of evolution is now in order.

CHAPTER SIX

The principles of evolution

In this chapter I outline a comprehensive theory of Darwinian and neo-Darwinian evolution that deals with the basic principles of the evolution of genes, phenotypes, individuals, populations, and species, but also extends into universal Darwinism as a set of principles that apply not only to competitions and interactions *between* species, but also to the adaptive resources *within* species and individuals.

Recent work in the field of evolutionary biology has moved well beyond the rather simplistic selfish-gene theory, which affords virtually all of the power of evolutionary change to competition between genes for self-replication, leaving human beings as little more than passive vehicles and gene carriers (compare Dennett, 1995, with Eldredge, 1995 and Wesson, 1994; see also Plotkin, 1994; Tooby & Cosmides, 1990b; Williams, 1985).

I therefore use *evolutionary epistemology*, especially the work of Plotkin (1994), as the main organizer of the principles of evolution developed here. In these terms, the essential evolutionary paradigm is a sequence of phases defined as: *variation, test or selection, differential reproduction, and the creation of fresh variants for a new round of testing and selection*. The principles that guide and constrain this *historical* sequence are understood to be relevant not only to the evolution of individuals, species, and populations, but to many other phenomena, including the fundamental nature of *current* adaptations themselves. I therefore offer a number of reasons to add the emotion-processing mind, our unit of evolutionary selection, to this auspicious list of within-organism entities and processes that operate according to these Darwinian principles of selection and favoured reproduction or descent.

The basic principles of evolution

There is a vast literature on the evolutionary issues pertinent to the union of psychoanalysis and evolution (see for example, Cosmides & Tooby, 1992; Dawkins, 1976b, 1983, 1987; de Duve, 1995; Dennett, 1995; Eigen, 1992; Gould, 1982, 1987, 1989; Ridley, 1985; Tooby & Cosmides, 1987, 1990a, 1990b, 1992; Wright, 1994). However, I have selected what appears to be the most comprehensive theory of evolution currently available, one that fits well with the broad issues I explore in terms of both the history of the emotion-processing mind and its present adaptive functioning. The centrepiece of this theory is Plotkin's (1994) masterful survey of the field in terms of *evolutionary epistemology* (see also, Campbell, 1974)—*the concept of adaptation as a mode of knowledge acquisition*. Plotkin's approach to evolution includes a conception of emotional adaptation that sees a hierarchy of contributions to the adaptive resources of *Homo sapiens sapiens*. Genetic influence is most basic, but also included are developmental factors and both individual and culturally shared intelligence.

Neo-Darwinism is a basic theory that attempts to account for the *transformations* of living, biological entities and structures, and their processes, in terms of the Darwinian principles of evolution. The following seems to characterize best the essentials of this science:

1. The theory of evolution is a statement of principles, regulari-

ties, and causes that properly describe the selective adaptive transformations or changes over time of individual organisms and their functional systems, doing so ultimately in terms of speciation and population changes—the creation of new species and the extinction of existing forms.

2. The key components of the theory of evolution are (see Plotkin, 1994):
   a. The existence of adaptive structures and their adaptations.
   b. The occurrence of *random or chance variations* in these adaptive structures, caused by such factors as mutations, genetic recombinations, and genetic drift—changes that are unrelated to, and therefore not directed, instructed, or shaped by, environmental factors.
   c. By means of competition between variants at the level of phenotype (the physical and mental expressions of underlying genes, the genotype), natural selection automatically and passively tests their comparative adaptive powers and contributions to survival (the ability to access and use energy sources—issues of economics) and reproductive fitness (sexual selection), choosing from these diverse forms those that are least costly, most efficient, and most successful in passing on their genes.
      i. The point of action of evolutionary forces is on phenotypes, which in turn leads to the selection of genotypes.
   d. The preferential reproduction and favoured descent of selected variants lead to new modes of adaptation. These variants then combine with fresh random variants to produce newly designed organisms, which then compete for survival with other existing variants.
3. The essential *process or mechanism* of Darwinian evolution is that of *natural selection as it leads to descent with modification*—competing chance variations acted on by natural selection, followed by the favoured reproduction and descent of the variations that best fit with and adapt to the existing environmental conditions, the fitness environment.
4. The *causes* of evolutionary descent are termed *selection pres-*

sures, and they are mostly found in an organism's fitness environment. Environmental change, an ever-present feature of biological life, is the motor for evolutionary change. Indeed, all evolution is *relational* in that it is based on the coupling of the organism with its environment.

5. Evolutionary processes are themselves subjected to change via evolutionary principles.

    a. It seems likely that the factors in the evolution of hominids, and especially *Homo sapiens sapiens*, are quite different from those that applied to earlier species. Human intelligence, culture, and social environments play a role in evolutionary processes to a greater extent than with prior species (Baldwin, 1896; Dennett, 1995; Plotkin, 1994). These emergent features of evolutionary change must be taken into account in developing an adaptationist program for the emotion-processing mind.

    b. Another unique feature of human evolution is that internally derived stimuli may serve as secondary selection pressures, especially those that are physical, repetitive, or spread out among existing populations.

    c. Inner psychological responses, conscious and unconscious, also may serve as weak secondary selection factors for humans in that human emotions can affect inter-organismic competition and intra-organismic states.

6. The fitness of an adaptation—its advantages over other existing or possible adaptations—generally involves a reduction in energy needs or costs and/or an increase in supplies of energy, and/or the enhancement of chances for survival and reproductive success, in part based on improved protection against predators and the greater availability of sources of energy.

7. Choosing the *units of selection* (Lewontin, 1970; Tooby & Cosmides, 1990b), the genetically founded *phenotypical* traits on which natural selection acts, is critical to a proper formulation of a Darwinian scenario. It is through the phenotype that the favoured reproduction of genes is effected. As I intend to show, for psychoanalysis, the emotion-processing mind appears to be an ideal entity or structure for this effort.

8. All living organisms are adapted to their respective environments—their fitness landscapes (Dennett, 1995; Eigen, 1992)—and they negotiate issues of survival and reproductive success within those partly stable yet ever-changing environmental conditions. The odds against species survival are great—some 98% to 99% of all of the species that have occupied this planet no longer exist.

9. In addition to the issues of survival and reproductive success, all organisms must negotiate a number of other basic issues common to all life forms (de Duve, 1995). The following are of note:

   a. autonomy–independence versus coalescence–dependence;
   b. stability versus instability;
   c. order versus disorder;
   d. simplicity versus complexity;
   e. symmetry versus asymmetry;
   f. instructionism versus selectionism.

   Choices in these areas made by organisms, based on unlearned instincts or on learning and intelligence, affect the organism's fitness and survival—and the survival of their offspring, as well as the evolutionary history of their species.

10. Propositions regarding the adaptive capabilities of *Homo sapiens sapiens* (and other species) must conform to two very different constraints from an evolutionary perspective:

    a. Nature and evolutionary processes tend to repeat themselves as fractals (that is, self-repeating and self-organizing configurations) or homologous designs and functions as organisms ascend the evolutionary scale. As a result of these *continuities*, many of the regularities and structural configurations governing and pertaining to the evolutionary histories and current adaptations of non-human species will apply in some form to human adaptations and their evolutionary histories as well (Kauffman, 1995).
    b. On the other hand, *Homo sapiens sapiens* is a species with features that are either absent in prior species or extant in only rudimentary fashion. Humans have the

exclusive capability of utilizing generative, spontaneous symbolic language (Bickerton, 1995). In addition, we possess a wide range of specific adaptive capabilities not seen in other species, including the ability to study ourselves. We have also created, and are therefore compelled to experience and cope with, social and cultural features and selection forces that are unprecedented in nature (Dennett, 1995).

These *emergent attributes and properties* of human beings and their adaptive armamentarium, as well as their distinctive fitness environments, speak for *discontinuities* and unprecedented adaptive capabilities seen in our species alone—whatever their earlier and fragmentary antecedents may have been. In developing an adaptationist programme for the emotion-processing mind, then, both continuities and discontinuities must be taken into account.

11. Because genetically based evolution is an extremely slow process, the genetically founded phenotypical configurations of present-day organisms including *Homo sapiens sapiens* were forged thousands of years ago in the ancestral Pleistocene era (Tooby & Cosmides, 1990b). It is therefore essential to identify the selection pressures and other pertinent factors that prevailed within this environment of evolutionary adaptedness. However, the existence of additional sources of current adaptations by means of naturally developed individual and shared intelligence (Plotkin, 1994) calls for an appreciation of current as well as past evolutionary issues and resources in generating a complete adaptationist programme (see below).

Knowledge, adaptation, and evolution

Plotkin (1994) and others (see Campbell, 1974) have made a strong argument for a hierarchy of evolutionary forces. Plotkin has argued that an organism's survival and adaptive capabilities depend primarily on genes and givens (instincts) that account for adaptations to slowly changing, long-term environmental events—conditions that change slowly enough to allow time for genetic alterations. However, other factors also account for evolutionary and adaptive change. These include epigenesis (development as it unfolds in a particular environment), individual cognitive structures and intelligence (phenotypes), and cultural or conspecific adaptations (socio-cultural considerations). These factors are especially critical to successful competitive adaptations in response to environmental changes that occur with some rapidity over the short term. Thus, while genes are constrained to respond to changes over the long term, individual and conjoint uses of intelligence enable humans and other organisms to respond to more sudden and unforeseen environmental happenings—uncertain futures (Plotkin, 1994; Waddington, 1969).

This approach to both evolution and adaptation proposes the existence of a nested hierarchy of factors with secondary control features. The ordering of levels is:

1. biological–genetic;
2. developmental;
3. cognitive learning (the acquisition by an organism of information about an aspect of its environment) and intelligence (which includes emotional cognition);
4. socio-cultural factors.

This layering of knowledge-acquisition mechanisms provides a wide perspective on the origins of human adaptations. Considerable stress has been placed on the so-called *environment of evolutionary adaptation*, the period during which adaptive mechanisms are selected and structuralized (Tooby & Cosmides, 1990b). On the genetic level, virtually all of these choices were made hundreds of thousands of years ago during the Pleistocene era, during which savannahs and other natural settings were the locales for hominid nomadic, hunter-gatherer species and their ways of life.

But this situation creates a likelihood that current genetically determined adaptations will fail to match present environmental conditions with any notable degree of effectiveness—genetic selection requires long periods of time to catch up with environmental conditions. However, the existence of additional sources of adaptive resources frees *Homo sapiens sapiens* from enslavement to the genetic factors in adaptation and allows for the use of human intelligence to generate better matches between contemporaneous environments and adaptive capabilities.

Universal Darwinism— the mind as a Darwin machine

Hierarchical layering is essential to Plotkin's (1994) work with evolutionary epistemology in which adaptations are seen as an organism's ways of knowing its environment and world. A central theorem in this approach states that adaptive entities that operate according to the general rules of learning, memory, intelligence, and cognition do so by adhering to principles that are comparable to the rules of evolution. Thus, our means of learning or acquiring knowledge of the environment, including aspects of emotional cognition and other mental/psychological functions (Gazzaniga, 1992; Jerne, 1955, 1967; Pinker, 1994; Pinker & Bloom, 1990), operate according to the so-called evolutionary analogy or *the principles of universal Darwinism—as Darwin machines* (Dawkins, 1983, 1987; Plotkin, 1994).

The essential points of this proposition are captured in what is termed the *g-t-r (generation–test–regeneration) heuristic* which defines the means by which an organism adapts to its environment in general, and to uncertain futures in particular. A heuristic is a strategy for solving a problem in an inventive manner, and the g–t–r heuristic is a model that is applicable to evolution itself, but also to all structures of adaptation that draw upon and follow evolutionary principles—it is a universal model of learning and adaptive change.

The elements of the g–t–r heuristic are:

1. the generation of variety, due largely to chance caused by mutations or variable environmental factors that occur during epigenesis;
2. a test phase, during which selection operates to effect the favoured reproduction of adaptively successful strategies;

3. regeneration of the favoured forms plus the introduction of new chance variants (and perhaps inventive, environmentally guided variants as well).

Heuristic strategies are hierarchically layered. The primary, fundamental heuristic is that of biological genetic development, which programmes the organism selectively to know and adapt to its environment. But there are also secondary heuristics that lie within the individual's phenotypes. The secondary heuristic systems include the immune system, a major healing system and our prime defence against microscopic predators; the human brain, the physical basis for human adaptation; the human mind, including its cognitive mental faculties like intelligence and language; and aspects of emotional cognition. Finally, there is a tertiary heuristic that stems from culture and the sharing of knowledge among individuals.

Contemporary Darwinian theory provides seminal and novel insights into historical, evolutionary processes. At the same time, the nature of the current adaptations that have resulted from these processes are clarified and afforded new understanding. These compelling ideas provide a sound foundation for the study of the unit of selection that dominates the rest of this book.

CHAPTER SEVEN

The unit of selection

The adaptive human mind and its emotional coping strategies have evolved over some six millions years of hominid existence. As noted, the first task in developing an evolutionary scenario is to choose and define a critical unit of evolutionary change. In the present context, this *unit of variation and selection* should possess sufficient power to illuminate and bring fresh thinking to human emotional adaptation and its issues (Lewontin, 1970).

I selected *the emotion-processing mind* for this purpose (see chapter 3). The choice was made empirically in light of clinical and formal science investigations, which indicated that:

1. The emotion-processing mind is a *definable entity* that can be modelled to yield unique insights into emotional adaptations.
2. The emotion-processing mind has a universal design, around which individual variations can be detected.
3. The emotion-processing mind is the source of emotionally charged communications, affects, and behaviours
4. This mental module is essential to human relatedness in that

it appears to be the basic unit of emotional cognition and adaptation.
5. The emotion-processing mind is a biological entity that operates according to fundamental laws of nature.

In support of this choice, Plotkin (1994) has stressed the equivalence between adaptation and knowledge acquisition and thereby established *cognitive capacities* as central to understanding the psychological adaptive resources of Homo sapiens sapiens. In this light, the selection of the mental module for *emotional cognition* as a basis for delineating the evolution of the mind seems apt.

Additional grounds for this choice are found in Donald's (1991) excellent efforts to trace the evolutionary history of general (non-emotional) cognition and language usage. His study of the stages of *cognitive evolutionary development* from early hominid life to the present was centred on an adaptationist program that deeply illuminated the nature of cognitive adaptive functioning. A comparable study of the cognitive capabilities of the emotion-processing mind promised to be similarly productive.

The broad scope and specificity of the emotion-processing mind provides a unit of selection that is more advantageous for study than the psychic defences and means of relatedness that were focused on by prior psychoanalytic evolutionists (Nesse, 1990b; Nesse & Lloyd, 1992; Slavin & Kriegman, 1992). Psychic defences are substructures of the emotion-processing mind and, as such, are not fundamental units. Relating is dependent on cognition. In addition, relating entails a complex set of behaviours that rely on so many deep structures and capacities that a scenario for the evolution of relationship structures over the aeons of hominid existence has yet to be accomplished.

The study of evolutionary factors in communicating and adaptive processing as carried out by the emotion-processing mind will both establish a critical evolutionary scenario of its own and offer a fresh basis for investigations into the development of relationship capacities.

Freud's views on the unconscious domain

To deepen our perspectives on the emotion-processing mind, I summarize briefly Freud's writings in this area (Freud, 1900a, 1923b; see also Langs, 1992c, 1995a). Freud (1900a, 1923b) offered two major psychological models of the mind:

1. His first model (Freud, 1900a) posited three systems of the mind—Ucs, Pcs, Cs—defined by the presence or absence of awareness of the contents in the system.

   a. The model was adaptively configured because it posited an external stimulus as the activator of the mental apparatus. It also envisioned transpositions from one system of the mind to another that vaguely suggested the possibility of processing mechanisms. Outputs from the psyche were defined in terms of behaviour and communication.

   Conflictual contents entered the mind consciously (Freud never specifically incorporated unconscious perception into his models of the mind—see Smith, 1991); once registered in awareness, they moved into the system Ucs, where they were in a state of, or under the influence of, repression. These conflicted contents could pass into the system Pcs by moving through a defensive censorship only if the contents were suitably disguised. These disguised Pcs contents, also outside awareness, easily entered the system Cs, where they attracted the hypercathexes of attention, although they did have to pass through a second and less intense censorship.

   b. This model of the psychic apparatus was *content-oriented*: The dynamics of the model depended on the nature of the psychical memories and fantasies at issue, and the qualities of the censorship or defences. The systems of the mind were seen as static entities, repositories for certain types of contents. They were not seen as processing systems, but as containers for non-conflictual and conflictual ideas, images, memories, and fantasies—although in the case of the systems Pcs and Cs, these troublesome contents were disguised.

c. Reality was acknowledged in this model, but the primary sources of neurosis were intrapsychic conflicts between the two main systems of the mind—Ucs and Pcs–Cs. These conflicts arose because of the intolerance of the system Cs for forbidden contents such as sexual and murderous oedipal wishes towards one's parents.

d. The primary and secondary processes posited by Freud (1900a) described the transformations, properties, and propensities of mental contents and did not reflect conscious or unconscious processing or adaptive mechanisms. Thus, the primary processes characterized Ucs contents, which were said to be drive-laden, to seek blind discharge or gratification, to disregard reality, to be easily displaced and condensed with other contents, and to utilize mobile cathexes. The system was able only to wish, and therefore its contents were not only without negations, they were timeless as well.

As for the systems Pcs and Cs, Freud (1900) claimed that the secondary processes that characterized these systems gave their contents a capacity for delay of discharge, used bound cathexes, kept them in contact with reality, afforded them a directness of expression, and involved a definitive time factor.

[In contrast to the Freudian model, the communicative model (see chapter 8) posits processing systems rather than contents and containers, and the deep unconscious system of the emotion-processing mind is seen as an intelligent system that is capable of delay, is highly perceptive, and is more in touch with reality than the conscious system. The latter is viewed as strongly defensive, as having poor contact with emotional realities, and as inclined towards action without suitable delay. In agreement with Freud, however, the deep unconscious system is viewed as expressing itself through narrative images created through displacement and condensation. This suggests that Freud was correct regarding the communication of unconscious contents but in error with respect to the nature of the mental processing systems responsible for these expressions.]

2. Freud's (1923b) second model of the mind was the structural

model, which posited an ego, id, and superego. The key points are:

a. The defining criteria for these systems were their *functions*. The use of conscious versus unconscious contents as defining features, or the possible use of conscious and unconscious attributes as a means of identifying these three systems, were no longer considered. In this new model, all three systems of the mind had conscious and unconscious components and functions. Thus, once Freud shifted to a processing model of the mind, he essentially discarded the guiding principle of conscious and unconscious functioning and communication as critical features of the psychic apparatus.

b. This model did little to encourage psychoanalytic evolutionary investigations. As for evolutionary research, the study of the vicissitudes of instinctual drives was left for ethologists and evolutionary biologists (Cronin, 1991), that of the superego to a variety of selfish-gene adherents (Wright, 1994), and that of the ego to cognitive psychologists and evolutionary psychoanalysts (Barkow, Cosmides, & Tooby, 1992; Cosmides & Tooby, 1992; Donald, 1991; Tooby & Cosmides, 1987, 1990a, 1990b, 1992). It is to be hoped that the communicative model of the emotion-processing mind will encourage more incisive psychoanalytic research in these areas.

Choosing the emotion-processing mind

I conclude this chapter by indicating some additional advantages to choosing the emotion-processing mind as the unit of evolutionary study:

1. The emotion-processing mind is a basic unit of emotionally related intelligence, knowledge acquisition, and adaptation—it is the basic organ for knowing and coping with the emotional world of human experience. It is an entity with structure and functions that has boundaries, an extended temporal duration, scale, and relative stability.

a. Entities of this kind universally are subject to evolutionary principles, both historically and with regard to their current adaptive functioning.
2. This unit of the mind accounts for adaptive processes that are both conscious and unconscious, thereby forming the basis for the investigation of specific forms of *unconscious adaptation*—an essential psychoanalytic pursuit.
3. The emotion-processing mind is a vital organ for relating and surviving environmental inputs on the one hand and, on the other, for the management of one's inner physical and mental state and self.
4. The emotion-processing mind generates quantifiable communicative expressions that have formed the basis for a formal, mathematically grounded science of psychoanalysis (Langs & Badalamenti, 1992a, 1994a, 1994b, in press). As such, it is a measurable, lawful entity and is therefore a biological product of a crucial evolutionary history.
5. The emotion-processing mind is a language-based structure. The spontaneous use of language and its symbols is the single most distinctive, evolved psychological structure and adaptive resource found in humans (Bickerton, 1990, 1995; Chomsky, 1980, 1988; Corballis, 1991; Liberman, 1991; Pinker, 1994; Pinker & Bloom, 1990).
6. The two-system model of the emotion-processing mind is more workable and parsimonious, and hierarchically more fundamental, than the existing structural model of the mind that postulates highly complex systems that are difficult to define and explore (Wilson & Gedo, 1993).
7. Clinical research has fathomed major aspects of the architecture of the emotion-processing mind and has unearthed several unexpected features of this mental module. These discoveries call for a careful exploration of the evolutionary forces that shaped the emotion-processing mind in this as yet unexplained manner.

A clear delineation of the structure and functions of this mental module will take us a step closer to the insights we are seeking.

CHAPTER EIGHT

The architecture of the emotion-processing mind

To illustrate the kinds of clinical observations on which the mapping of the design of the emotion-processing mind is based, I introduce a brief clinical vignette and then add some insights derived from the other hierarchical levels of psychoanalytic observation and formulation before finally turning to a proposed evolutionary scenario for this mental module.

Ms Wells*, a woman in her early 30s, had sought treatment with Dr March, a male psychotherapist, for a severe depression that had developed after her father was killed in an automobile accident. She was referred to Dr March by her friend, Alice, who was also in therapy with him. Dr March's office is attached to his home, and his wife and two young

---

*As has been my practice for some time, all of the clinical material in this book is fictitious. The vignettes are offered solely as models and narrative illustrations. They are, however, faithful to reality as observed from the communicative vantage-point.

sons are quite visible to his patients, much as his patients are visible to them.

Some three months into her once-weekly therapy, Ms Wells arrived for her session and found another woman in the waiting-room. They both sat there waiting, but Dr March failed to appear at 5 p.m., the time Ms Wells's session was to begin. Ten minutes later, he entered the waiting room and was visibly surprised to see Ms Wells sitting there. He sent the other woman into his consultation room and then asked Ms Wells why she was there. She responded by reminding him that it was Thursday and time for her session, and he laughed and revealed that he had had a lapse and thought her session was on Friday; he had scheduled another patient in her time. He would be free at 6 p.m. and would gladly delay his dinner to see Ms Wells then, if she would come back in 45 minutes. Ms Wells agreed to do so.

When she returned for her belated session, Dr March immediately apologized for his error, explaining that he seemed to be rather forgetful these days. Ms Wells said she realized that he had a very busy practice and added that anyone can make a mistake; besides, seeing him an hour later wasn't any big deal. She went on to say that she had sat in her car and had dozed off. She had had a brief dream in which her sister, Kim, was seated at the kitchen table in the house in which they had been raised, typing at a computer. A dark-looking woman came into the kitchen, yelling that the house was on fire, at which point Ms Wells woke up.

Associating to the dream, Ms Wells noted that the woman who was in the waiting room earlier that evening was dark-complexioned, and so is Alice, her friend who also is in therapy with Dr March. Ms Wells had spoken to Alice the previous day and, well, they both had remarked that Dr March was a really good therapist; he seemed to know how to get at the heart of things like no other therapist either of them had seen before. Dr March reminded Ms Wells of her father—he, too, was very bright and quite absent-minded.

Dr March responded to this material by noting that a father transference seemed to be in evidence, and he asked Ms

Wells to associate further to her dream. Her next response was that she could see what Dr March meant—she had mentioned her father, hadn't she? She felt rather stupid for not having seen it. If she keeps talking, she'll reveal even more of her ignorance; it's better to be silent.

After a pause, Ms Wells went on to say that her sister Kim was now living with a man and another woman who, like Kim, is his girlfriend. It's the kind of situation that Ms Wells would find intolerable. The idea of sharing a man with another woman is not only repugnant, but masochistic and immoral; it gives license to the man to exploit and use not one but two women. Her sister is blindly letting herself be the victim of this reckless man. Ms Wells wouldn't stay in a situation like that for ten minutes. She's furious that Kim won't listen to her when she warns her sister that she's being used and abused and that she will end up badly damaged by it all.

The fire recalled a memory of Ms Wells's father, who was an accountant who had his office at home. His clients would come to the office and when she was in her teens, and some of them would flirt with Ms Wells, who hated their cheap remarks and lurid looks. There had been considerable tension between her mother and father during those years, and Ms Wells had had the impression that it had to do with her mother accusing her father of having had an affair with one of his clients—which turned out to be the case. Ms Wells can still remember her mother crying and complaining that this woman had taken her place in bed; her mother was enraged, yet somehow she didn't do anything about the situation. She should have stood up to Ms Wells's father and ended the farce by forcing him to choose between herself and the other woman once and for all.

On one occasion, there was a short circuit in the electrical wiring in her father's office, and the house had caught fire, causing a great deal of damage. It turned into a nightmare when they discovered that their maid had been felled by the smoke and trapped in the kitchen; she had succumbed to smoke inhalation. Soon after, the maid's sister had come to work for them; it was eerie, as if the maid had never died.

Formulating the material

To formulate the transactions of this session on the basis of the communicative approach, I use an adaptationally oriented framework for assessing the material, supplemented by psychodynamics, personal genetics, and other necessary viewpoints. The entire effort is shaped by the overriding goal of using this vignette to afford us a sense of the architecture of the emotion-processing mind so that we can apply evolutionary theory to our chosen unit of selection.

The first task is to clarify the status of the conditions and ground rules of this therapy by reviewing the basic patient–therapist contract. As noted, adherence to the configuration of *unconsciously validated* ground rules constitutes *secured-frame therapy*, while any departures from these unconsciously preferred ideals create a *deviant or frame-modified treatment setting*. Clinical observations show that each set of framework conditions—secured or not secured (modified)—has a distinctive and compelling set of meanings, implications, and effects on both parties to therapy.

In this case material, there are two notable basic frame alterations. The first is Dr March's home office, which modifies the therapist's relative anonymity (both his house and his family are revealed to the patient) and the confidentiality of the therapy (the patient is seen by the therapist's family members). The second alteration is the referral of the patient to Dr March by one of his other patients, which is a modification of the one-to-one relationship and of the total privacy of the therapy. As her dream and other associations suggest, even though the two friends do not share sessions, the unconscious experience is that each is present in the other's sessions. These are, then, the basic modifications of this therapeutic contract or frame, and they will be kept in mind as we proceed.

The session itself is subjected to an immediate frame modification of the ground rule regarding the time of Ms Wells's sessions. Dr March elected to change the hour because he had scheduled another patient in Ms Wells's time. The presence of the other patient constitutes an additional frame break in that it brings a third party into Ms Wells's therapy. This compromises the total confidentiality and privacy of her treatment. Ms Wells's

82 SOME BIOLOGICAL PRINCIPLES

material must therefore be formulated in light of these powerful adaptation-evoking, frame-modifying trigger events.

Ms Wells begins her session with some direct comments on the time change, excuses her therapist for his lapse, minimizes its importance, and goes on to praise her therapist for his work with her and her friend and fellow patient, Alice. In so doing, Ms Wells has manifestly alluded to (represented) two frame-deviant, adaptation-evoking triggers—the change in the time of the session and the patient referral—and she has treated both frame-modifying interventions as relatively insignificant.

The conscious response

Let us pause to examine and formulate Ms Wells's *conscious adap-tive reactions* to the most immediate, identified trigger events connected with this session as revealed in her direct remarks. We should note the following:

1. This situation involves a consciously recognized trigger event (Dr March's lapse) and a direct, conscious adaptive response—mainly in the form of Ms Wells's acceptance of the postponement of her hour and forgiving her therapist for his lapse.
2. An objective evaluation of this trigger would mark it as hurtful, insensitive, rejecting of Ms Wells, or even manipulative. This assessment runs counter to the patient's conscious forgiving response.
3. In this light, Ms. Wells's conscious reaction to the trigger appears to be highly defensive and denial-based.
4. This suggests that the conscious mind—*the conscious system*—is a defensively organized system of the mind.
5. In terms of evolutionary processes, most biological mental systems have evolved to enhance an organism's (conscious) perceptual abilities and capacities to learn about the world. In contrast, it seems that the human conscious system has evolved with tendencies to reduce and screen out both per-

ceptions of and knowledge about the environment of immediate adaptation and about specific anxiety-evoking external events.

6. Ms Wells's manifest communications and adaptations and her conscious, direct statements minimize the importance of this frame deviation and offer praise and support to her therapist despite his hurtful error. Adaptively, the patient was able consciously to quickly dismiss this deviant frame issue and move on to other matters that were *evidently* of greater concern to her. Were it not for unconscious processing, there would be little more to say.

    a. This assessment is based on manifest behaviours and direct communications, and it reflects the kind of formulation of these transactions that are developed by therapists who exclusively attend to the manifest contents of their patients' material—and their evident implications. As a result, their view of the emotion-processing mind is essentially a conscious-system view—what I call a unidimensional, flatland view of the human psyche (Abbott, 1884; Langs, 1992c).

On the basis of observing and formulating the manifest contents/evident implications of the patient's material, the human mind seems relatively unaffected by violations of the ground rules of psychotherapy. From the usual psychoanalytic perspective, the primary issue facing this patient is likely to be construed as the activation, largely from within herself or possibly from a highly prejudiced experience of the therapist's activities, of a so-called father transference—essentially, that Ms Wells was confusing her therapist with her father. The role of the therapist's frame-related behaviours are relatively inconsequential for this formulation, and on this basis there would be little possibility of realizing that the emotion-processing mind operates not only with a conscious system, but with a second adaptive system as well. Its full architecture would go unappreciated.

Defining the unconscious domain

A second type of formulation goes beyond the patient's surface responses and is crafted through *trigger-decoding*. This method is experience- and theory-based. Its goal is to assess this patient's world of deep unconscious experience and the encoded meanings in her material. As we will see—and this point is appreciated by scientists in other fields—the means by which one observes nature, and organizes and formulates the meanings of the resultant observations, essentially determines the theory and conclusions that one generates. All of science is dialectical and interactional, an interplay in never-ending cycles of theory, observation, formulation, and revision of theory.

The primary province of psychoanalysis is the domain of *unconscious* communication, processes, and adaptations. As early as 1900, Freud specifically introduced the concept that a patient's free associations are laden with both manifest and latent contents and meanings. The adaptation-oriented, communicative approach has refined this vital insight in the following ways:

1. *Narrative or storied associations*, in the form of dreams, daydreams, personal recollections of events and happenings, and similar modes of imagistic and thematic expression, serve as the principal carriers of significant latent contents (Langs, 1992a, 1994b, in press).
2. There are two very different types of latent contents:
   a. the unstated implications of a manifest message;
   b. encoded meanings, which are disguised within but cannot be discerned from a reading of the direct message, although they can be abstracted and decoded in light of their adaptation-evoking trigger events.
3. The world of unconscious experience, as it pertains adaptively to *repressed trigger events* and to the *repressed meanings of known (conscious, non-repressed) trigger events*, is revealed by decoding the disguised themes in a patient's narratives in light of their evocative triggers.
4. The encoded meanings of a patient's free associations are, then, disguised reports on the patient's unconscious percep-

tions and processing of the therapist's interventions—behaviours, verbalizations, affects, and frame-management efforts.
5. To ascertain these encoded meanings, *trigger-decoding* is necessary. This is done by identifying the key adaptation-evoking trigger events for the patient's narrative communications and transposing the displaced themes in these storied images from their manifest story back to the repressed trigger event.
   a. For psychotherapy patients, frame-related interventions by their therapists are *unconsciously* the most powerful class of adaptation-evoking triggers.
   b. Because the experience of the meanings of these triggers occurs largely outside awareness, patients typically invoke the mechanisms of *repression and displacement* in responding to these events. Thus, a patient responds to a frame-related trigger event in therapy with a story about someone other than the therapist and about an event other than the frame impingement. This kind of communication reflects the patient's automatic (unconscious) way of disguising or encoding unconscious perceptions of trigger events whose meanings are consciously intolerable. These meanings are perceived subliminally and then both experienced and processed unconsciously.

   The encoded images from a patient therefore symbolically convey both the nature of an unconscious experience and the results of the patient's unconscious intelligent, adaptive processing of its subliminally registered meanings. These unconscious reactions typically include encoded recommendations from the patient to the therapist as to the trigger event at hand. These directives usually pertain to securing modified aspects of a treatment frame and/or to intervening more correctly when a therapist has been in error.
   c. The unconscious sequence can be outlined as:
   trigger event → subliminal perception → unconscious processing → encoded adaptive response → conscious, but *encoded*, narrative.
6. Simply put, deep unconscious experience is the level at which the power of emotional events is expressed and at which the latter exert their most telling effects.

Possible arrangements of the emotion-processing mind

In the clinical vignette there is a consciously acknowledged trigger event, and we should expect to find unconscious responses to its *repressed* meanings. To clarify this point, I list what appear to be the three possible types of responses to trigger events:

1. Trigger events might be dismissed both consciously and unconsciously—that is, both systems of the mind might experience a given trigger event or environmental impingement as relatively inconsequential.
2. Trigger events might evoke definitive reactions from both systems of the mind in forms that are similar and complementary to each other. This would be an indication that the two systems of the emotion-processing mind are aligned with each other and view the world in comparable fashion.
3. The conscious and unconscious responses to trigger events might be dramatically different. This would mean that the assessments of trigger events by each system are disparate and quite distinctive. It would also imply that one system experiences meaning, while the other does not—that is, one system is more defensive than the other system. Our preliminary analysis suggests this may be the case and that it is the conscious system that adopts the more defensive posture.

In addition to dealing with *consciously acknowledged* trigger events, a patient will also respond to one or more trigger events that have been *repressed* (some of them permanently so, if no intervention intercedes). This means that, via condensation and encoding (symbolizing), the same narrative associations can deal with both known and unknown triggers. As a rule, this arises when the conscious system has repressed a particular trigger event that the deep unconscious system has perceived unconsciously and is processing. In this case, then, the two systems of the emotion-processing mind would be operating with very different views of the emotional world and its triggering events—and of people, places, and events.

Whenever a patient reacts to aspects of a trigger event without conscious awareness, the response cannot be reflected in direct comments about the trigger (these would, by definition, be part of a conscious response). Instead, the patient automatically turns to stories about situations other than the recent intervention of the therapist. Almost always, these allusions refer to events and incidents that have occurred outside of the treatment setting (more rarely, a patient will use a story about one incident in therapy to encode an unconscious reaction to another, contemporaneous incident or trigger).

Let us look now at how Ms Wells responded unconsciously to the two frame-modifying trigger events she had experienced just minutes earlier, in order to see which of the three above-noted possibilities is actually in evidence.

Trigger-decoding Ms Wells's material

Ms Wells manifestly dropped the subject of her therapist's frame lapse, but we will examine her subsequent associations to determine whether they support the communicative expectation that while the conscious and direct material no longer dealt with this problem, her latent, encoded material continued to do so. Her manifest dream—dreamt after the incident in the waiting-room—is a narrative about the patient's sister and a strange woman who yells that the house is on fire.

To establish the connection between the surface dream and the frame modifications, we look for *bridging imagery—themes shared by the meanings of these frame-deviant triggers and the manifest dream stories and images.*

Thus, the *intruding* woman in the dream can readily be taken to represent the woman who *intruded* into the waiting room and into Ms Wells's hour. But in addition, through *condensation*— the capacity of dreams and other narratives to represent simultaneously two or more reactions to a given trigger event and/or reactions to two or more trigger events—the woman may also represent Ms Wells herself *intruding* into her friend Alice's therapy space (the reverse, the friend's *intrusion* into Ms Wells's space, is possible but less in keeping with what actually had happened). Another bridging theme can be found in the allusion to Ms Wells's house (and later, to her father's home–office arrangement), which can be taken to represent Dr March's home–office setting.

In this light, the fire seems to portray the dangerous qualities of these frame deviations, but we need more imagery to develop a fuller sense of the patient's unconscious experience of these frame violations. For the moment, however, these formulations do find support in Ms Wells's first associations to her manifest dream, in which she indicated that the strange woman in the dream looked like both the woman who was sitting in Dr March's waiting room and Alice, her friend who also is in therapy with Dr March.

In all, these *bridging themes* support the thesis that in most instances when a patient is dealing with an acute frame-modifying event, the subject is manifestly dropped but is followed by a shift to narrative communications whose *latent contents encode his or her unconscious responses to the manifestly dismissed trigger event.*

There are, in all, at least six frame-deviant, adaptation-evoking trigger events that Ms Wells seems to be dealing with at this point in her session:

1. the presence of the woman in the waiting-room;
2. Dr March's not seeing Ms Wells at her appointed time;
3. his seeing the other woman during the time allotted for Ms Wells's session;
4. his seeing Ms Wells at a time that is different from that of her scheduled session;
5. the acceptance by Dr March of Ms Wells as a referral from another patient and the continuance of the referring patient's psychotherapy with him;
6. the home–office setting.

Returning to the session, the next story the patient tells is about her sister, Kim, and the narrative introduces the *theme* of a *ménage-à-trois*—two women living with the same man. We are therefore obliged to find the trigger events that are represented by this theme and link this encoded or disguised image and its themes and meanings to these triggers in order to reveal the patient's unconscious experience of the events in question.

While manifest images are single-meaning vehicles of communication, encoded narratives carry multiple meanings in two senses: (1) narratives have two distinctive levels of meaning—manifest and latent/encoded, and (2) the encoded meanings themselves are multiple in nature. The latter is achieved by the mechanism of *condensation* as it operates in creating disguised images. The critical point, however, is that the *multiple meanings that are condensed into a single narrative element are constituted as encoded adaptive responses to two or more trigger events*. Thus, a given dream or story will condense into a single image unconscious adaptive responses to two or more trigger events that are active for a patient at any given moment.

For example, by invoking *bridging themes*, we can argue that this first story encodes Ms Wells's unconscious perceptions of all but one of the six disruptive frame experiences to which she was currently adapting unconsciously. This is the case even though, consciously, she has not as yet alluded to several of these triggers and she has stopped responding directly to the two triggers to which she had initially directly referred. Only the change in the time of Ms Wells's session lacks an evident bridging theme for the moment.

Ms Wells's images indicate that she unconsciously experienced Dr March's frame-modifying interventions as creating a hurtful, exploitative threesome, and, in addition, that her own conscious acquiescence to this arrangement was self-hurtful and masochistic. Through condensation, these images allude to the triggers of the appearance of the woman in the waiting-room (the therapist must bear full responsibility for this third-party frame alteration); the home–office setting (Dr March's bringing his wife into his relationship with Ms Wells); and Alice's presence in therapy with Dr March.

Remarkably, then, the emotion-processing mind is able to create multiple simultaneous layers of representation of meaning. The adaptive resourcefulness of human language and its symbols—and, through them, of the emotion-processing mind—is multi-levelled, diverse, and powerful.

Language acquisition affords the human mind a variety of conscious aptitudes for thought and action. But language also enables the human mind unconsciously to work over, to represent, and to solve a multitude of adaptive problems with a single responsive image. The advantages in adaptive resourcefulness and intelligence over the best levels of animal intelligence are immense. Compared to the quite constricted conscious mind, the deep unconscious mind seems unbelievably expansive and incisive.

As for encoded meanings and adaptive solutions, Ms Wells was quick to state that she saw this kind of threesome as immoral (rule-breaking), untenable, abusive, and self-harmful. She then added that she would not tolerate an arrangement of that kind for ten minutes. This comment is a *model of rectification—an unconsciously arrived-at adaptive recommendation.* Through her encoded imagery and comments, Ms Wells is telling her therapist and herself that the compromised arrangements of, and recent frame breaks in, her therapy are exploitative and untenable, and she should leave her therapist. She also makes an unconscious self-interpretation that *not* to do so is to be masochistic. And yet, despite the strength and clarity of these encoded recommendations, Ms Wells (consciously) remains in therapy with Dr March. This raises the obvious question: Why is this so?

Clues to the emotion-processing mind

To begin to answer this question, we need the following additional perspectives:

1. *Consciously*, Ms Wells had accepted or even sought out a psychotherapy that was constituted under basically deviant-frame conditions. She did not evince a conscious problem with seeing Dr March in his home–office or with being referred to him by one of his patients, while his lapse on the day of her session caused hardly caused a ripple of a conscious protest on her part.

a. *Ms Wells's manifest-conscious messages are consonant with her actual adaptive–responsive behaviour*—that is, acceptance of the situation and its many deviations.
2. Ms Wells's *unconscious* responses to these same triggers are almost diametrically opposite to her conscious reaction. This opposition is an indication of a split in the emotion-processing mind, which reveals a dramatic discontinuity between conscious and unconscious adaptive reactions to trigger events—at bottom, between the conscious and deep unconscious systems of the emotion-processing mind.
   a. Instead of seeing these frame breaks as innocuous or irrelevant (the conscious view), *unconsciously* Ms Wells perceived and experienced them as inappropriately seductive, abusive, immoral, and intolerable. And instead of simply continuing with her therapy (her conscious choice), Ms. Wells unconsciously advised herself to get out of treatment.

These observations provide us with a first answer to the question of why Ms Wells behaved as she did with Dr March. It appears that our deeper wisdom and stronger adaptive resources in the emotional realm are unconscious rather than conscious. But the design of the emotion-processing mind is such that these effective unconscious coping capabilities do *not* influence direct, conscious adaptive efforts. As humans, we are cut off from our own deeper knowledge. And as a result, as we saw with Ms Wells, we typically engage in self-hurtful behaviours and make self-harmful conscious choices, even though we know better unconsciously. Discovering why the emotion-processing mind has evolved in this manner is a basic challenge for the adaptationist programme that I propose here.

Some design features

To clarify the problem at hand, I will summarize what Ms Wells's rather typical set of responses reveals about the architecture of the emotion-processing mind:

1. The conscious system and the deep unconscious system of the emotion-processing mind are separate adaptive systems, discontinuous rather than continuous.
2. Information and meaning do not flow directly and without disguise from the deep unconscious system to the conscious system. There are, then, no break-throughs into direct awareness of unconscious perceptions and the unconscious processing of their meanings.
3. The only communicative contact point between the two systems of the mind is via the encoded messages emitted by the deep unconscious system, which are expressed in manifest narratives only through disguised images.
4. The conscious system is activated by *conscious perceptions*, while the deep unconscious system is activated by *subliminal or unconscious perceptions*. Thus, *the separateness of these two systems begins at the perceptual level and extends from there.*
5. The mode of processing to which a given event and each of its essential meanings are assigned is determined at the perceptual level quite automatically and without awareness. Furthermore, the selected mode of perception—*whether conscious or unconscious*—is fateful for the subsequent processing of each distinctive meaning of an emotionally charged event.
    a. This selectivity applies to the trigger event itself, which may or may not register in awareness in its entirety. A large number of emotionally charged, traumatic trigger events created by psychotherapists suffer from the fate of total denial and repression by their patients. This principle also applies to the outside lives of both parties to therapy.
    b. The same selectivity also applies to the various meanings and implications of a consciously recognized trigger event. These meanings are automatically separated into those that are allowed entry into awareness and those that are not.
    c. The factors in this assignment process are twofold:
        i. the nature and meanings of the traumatic event (the therapist's contribution);

ii. the sensitivities of the recipient, including his or her modes of processing incoming information and meaning (the patient's contribution).
  d. As noted, the conscious registration of a meaning of a trigger event is followed by conscious adaptive processing, while unconscious registration is followed by unconscious adaptive processing.
6. The two processing systems of the emotion-processing mind operate in tandem, as parallel processing systems of adaptation. However, they do so quite independently and in response to very different meanings of a given trigger event—that is, to very different aspects of human experience.
7. Each system has its own perceptual range; values, needs, and motives; ego, superego, and id subsystems; primary concerns and issues; and intelligence, adaptive resources, memory system, and preferred coping strategies.
8. The deep unconscious system's intelligence and adaptive capabilities in the emotional domain appear to be far superior to those of the conscious system, which seems to respond to environmental impingements with defensiveness, maladaptive reactions, and choices that are ineffectual and self-hurtful.
9. The conscious system is a relatively effective *survival system*. However, it appears to be far less capable of adapting well to emotionally charged impingements.
10. The conscious system appears to be under the unwitting influence of *unconscious needs for self-harm*. This indicates that in addition to its unconscious processing wisdom subsystem, the deep unconscious system has another subsystem—*the fear–guilt subsystem* (Langs, 1993a, 1995a). This latter subsystem embodies unconscious needs and motives related to human fears of personal death and to guilt that is related to wishes for punishment and self-harm for transgressions real and imagined.
11. The only known outputs available to the processing efforts of the deep unconscious wisdom subsystem are *communicative* in nature, and they are consistently subjected to encoding and disguise.

94   SOME BIOLOGICAL PRINCIPLES

  a. This subsystem's processing efforts and its resultant adaptive choices do *not* affect conscious reasoning and behavioural responses, largely because deep insight is never revealed in direct form and therefore is unavailable to awareness. The deep unconscious wisdom subsystem also exerts few if any silent effects on conscious or direct adaptation. Intuitive adaptive responses do not arise from this system, but emanate instead from the superficial unconscious subsystem of the conscious mind.
12. However, in contrast to the lack of influence from the deep unconscious wisdom subsystem, deep unconscious fear and guilt appear to have considerable silent power over conscious adaptations—a treacherous and endangering, unexpected design feature of the emotion-processing mind that has major consequences for both psychotherapy and emotional life in general.
13. The overall perceptual range of the conscious system is wide and far-reaching; the system is sensitive to a great variety of impingements from without and within. However, the conscious system is, in general, relatively insensitive to impingements related to rules, frames, and boundaries, and it tends to prefer, and to be very tolerant of, frame alterations.
  a. Conscious thinking is especially inclined towards psychodynamic and personal genetic issues, and less able to negotiate and grasp the other levels of human experience—especially those related to rules, frames, and boundaries.
14. The deep unconscious system is focused sharply on settings, rules, frames, and boundaries. The system advocates secured frames as reflected in its emitted encoded narratives and consistently validates their presence or invocation. The deep unconscious system is also highly sensitive to and critical of the hurtful qualities of frame alterations (attributes that typically are denied consciously).
15. With respect to adapting to the emotionally relevant environment, the deep unconscious system is far more open, sensitive, and effective—yet its voice is all but unheard consciously. Instead, conscious adaptations fraught with maladaptive features are the predominant direct causes of

behaviour, although they are, of course, driven by disturbing unconscious perceptions and motives of which the patient is quite unaware.

16. Finally, these observations and postulates imply that *displacement, denial, and repression are inherent to the design of the emotion-processing mind*—they are invoked repeatedly and automatically by the human mind at times of stress. The activation of these mechanisms is, of course, governed by the psychodynamic and personal genetic qualities and meanings of a given stimulus or trigger event. That is, the emotion-processing mind is designed for the universal use of displacement and repression, but the meanings that are displaced and repressed are matters of individual unconscious selection on the basis of genetics, developmental incidents, and life experiences. Still, design (psychoanatomy) is a more basic hierarchical level than psychodynamics, and, as a result, in the presence of an emotional dysfunction, the architecture of the emotion-processing mind is far more powerful causally than psychodynamics—and also more difficult to modify.

    a. These propositions are relevant to clinical practice in that patients consistently deny, repress, and displace their reactions to their therapists' interventions to figures outside of therapy—far more than the reverse (their making displacements from outside figures onto the therapist—so-called transferences). This point has been verified clinically and is also in keeping with the fundamental evolutionary principle that organisms are designed to adapt to the immediate external situations in which they find themselves. It also reflects the powerful mantle unconsciously afforded to psychotherapists as healers and descendants from the shamans and priests—their interventions are greatly idealized and invested in unconsciously by their patients.

    b. For therapists, the main disruptive consequences of the automatic and universal (inescapable) use of these three defence mechanisms comes from displacements that move from their patients to their outside lives—although they also move from their outside lives to their patients as well. This latter type of displacement is inevitable and extremely common, though seldom appreciated as such. Nevertheless, the selection of behaviours and interventions made by therapists with each of their patients is deeply affected by trigger events and unconscious experiences in their social and professional lives. A great deal of self-monitoring and trigger-decoding is needed to reduce these inappropriate displacements to their utmost minimum.

Knowledge reduction

Clinical evidence indicates that the *deep unconscious system* is relatively *non-defensive* in accumulating knowledge about the environment, while the conscious system is quite the opposite—entrenched with defensive devices and obliterating mechanisms. *Thus, the evolved design of the emotion-processing mind suggests that in some fashion both survival and reproductive fitness were and are served by nature's selecting for the reduction of conscious knowledge of the enormously complex emotionally charged environments faced by humans.*

While naively we would have expected the emotion-processing mind to have evolved to enhance conscious contact with the environment through knowledge acquisition via cognitive–mental–intellectual–memory means, this does not appear to have been the choice made by natural selection. Indeed, the trade-off as reflected in cost–benefit ratios seems to involve pitting the great cost of knowledge reduction against the price that would be paid for not protecting the conscious system from the overload that would result from the full awareness of emotionally charged perceptions and their very disturbing meanings and implications (see chapter 11).

This knowledge-reduction mechanism appears to have a rather broad biological forerunner in the highly selective attributes of cognition and memory seen in all species. No organism responds to all levels and all types of environmental encroachments; selectivity and parsimony are the rule. Animals have evolved so that they are wired to know exactly what they must learn in order to survive and reproduce, and the selectivity of this search for knowledge depends greatly on the nature of the fitness environments in which they have evolved. This concentration evidently enhances adaptation by promoting a focusing down on critical information and meaning and a reduction of unneeded distractions.

Nevertheless, the cost of this kind of narrowing of pre-human perception seems small compared to its present version in the minds of *Homo sapiens sapiens* and the extreme extent of the reduction in contact with the external world that has been engineered into the conscious system of the emotion-processing mind. But then again, the adaptive problems and selection pressures that have led to the development of this mental module also were extraordinarily complex, quite overwhelming, and in many ways entirely unprecedented. We are dealing with a unique, emergent moment in the history of biological species.

In all, these unexpected design features will need to be taken into account and explained when we develop our adaptationist programme for the emotion-processing mind and explore this module as a possible Darwin machine. But, for now, let us return to our clinical vignette to see what further analysis of the material can add to our picture of this mental module.

Some additional formulations

Ms Wells next associated to the fire, which recalled her father's home–office arrangement. As noted, this bridging theme is a clear, displaced representation of Dr March's home–office setting. In principle, an immediate frame deviation will tend to activate background frame deviations, which then find further communicative expression and evoke belated adaptive reactions as reflected in the material at hand.

The home–office recollection brought to mind Ms Wells's being subjected to seductive actions by her father's clients. Through condensation, this is an encoded allusion to the intrusion of another patient into the therapist's office at the time of Ms Wells's session, and to the seductive contaminations caused by the therapist's home–office frame alteration (although not referred to as yet in the session, while she was waiting in her car for her session to begin, Ms Wells had seen Dr March's wife going into their house). In responding *unconsciously* to these frame-deviant experiences, the patient voiced her objections to this arrangement, though here, too, consciously she accepted it without complaint or manifest comment.

The seductive and frame-modifying aspects of this collection of third-party deviations are restated in the story that follows, in which Ms Wells recalls that her mother had accused her father of having an affair with a client, which proved to be true. Again, through condensation, this image undoubtedly alludes to Ms Wells's unconscious experience of Dr March's involvement with the woman who took Ms Wells's appointment; to seeing his wife enter their home; and to the presence of her friend Alice in her therapeutic space. In one way or another, each of these women symbolically but meaningfully has taken Ms Wells's place in bed with her seductive therapist.

Threesomes, infidelity, betrayal, and frame breaks abound in highly traumatic forms in this material. Yet Ms Wells raises nary a conscious protest about how she is being treated by her therapist. She continues to idolize him, while evidently suffering unconsciously in ways that are certain to affect her behaviour and life adversely. This again speaks for apparent and unexplained peculiarities and inefficiencies in the present design of the emotion-processing mind.

The material continues with the recollection of a fire that had started in her father's office, trapping the maid who died and who was replaced by her sister. This encoded imagery appears to reflect the destructiveness and entrapping qualities of the many frame breaks Ms Wells was experiencing. But, in addition, it also reveals the main unconscious reason why Ms Wells chose Dr March as her therapist and had stayed with him despite his hurtful and damaging interventions.

One of the strongest unconscious motives for frame alterations involves their use to deny one's mortality and to cope, in however costly a manner, with a major loss, especially through the death of a loved one. This is particularly clear with patients who refer friends or relatives to their therapists and with the patients who accept or seek such referrals. The presence of the third party creates a situation in which possible loss is, in fantasy, magically avoided—if one party dies, the other two survive and continue on.

Unconsciously, by this means, death is denied and death anxiety assuaged—the situation creates an unconscious delusional belief in one's immunity to loss for both patient and therapist. But the cost-benefit ratio is high because the defence is essentially maladaptive; sustaining it requires life-long frame breaking and the use of manic defences that typically are self-harmful and harmful to others. Considerable damage is done by the deviations involved, and a pattern is set for patients who have suffered a loss through death to live their lives enacting denial-based defensive behaviours. As a result, they find themselves forever searching for inappropriate replacement people whom they need always by their sides, lest their death anxieties, which are not in any way worked through and resolved by this means, overwhelm them.

In addition, the harmful aspects of the frame alterations satisfy the patient's needs for punishment caused by the deeply unconscious survivor's guilt from which they suffer. The net result is a maladaptive life of self-harm and inappropriate relationships, of which the therapist becomes a notable instance—the therapist plays into this maladaptive pattern all too well. Unconsciously, Ms Wells had selected Dr March largely because of, rather than despite of, his frame alterations—it was her way of denying and maladaptively coping with the death of her father.

The second segment of this vignette has amply confirmed our initial formulations regarding the architecture of the emotion-processing mind. In all, we have been confronted with a number of unexpected and seemingly inefficient and self-defeating aspects to the adaptive design of the human mind. They materialize for view only through adaptive-interactional listening and formulating, and through attention to unconscious, encoded communications and adaptive responses to environmental impingements. With our challenge now well defined, it is time to prepare ourselves specifically for our venture into evolutionary history.

PART III

EVOLUTIONARY SCENARIOS FOR THE EMOTION-PROCESSING MIND

CHAPTER NINE

Setting the stage for an adaptationist programme

We turn now to the first of our two main challenges in evolutionary psychoanalysis, the delineation of an adaptationist programme for the emotion-processing mind. Guided by the principles of Darwinian evolution, I will develop this programme to deal with the following issues:

1. The key features of the present design of the emotion-processing mind, our unit of selection.
2. The origins of, and possible random variations in, the design of this mental module.
3. The selection pressures that interacted with the genetic foundation of the structure of the emotion-processing mind, shaping and constraining the choices made through natural selection.
4. An overall explanation of how the present design of the mind came into being.
5. A scenario for the possible future developments of this critical module of human adaptation.
6. A better understanding of the process of emotional adaptation, with a special eye towards how the insights yielded by

this means illuminate both psychoanalytic theory and the therapeutic process.

Planning the scenario

We have a lot to manage. We need to understand the best available principles and guidelines for creating an adaptationist programme and to fashion a scenario that will deeply clarify the many puzzles we have identified with respect to the design of the emotion-processing mind. We also need a broad and informative set of evolutionary principles to inform this work, as well as a knowledge of the history of the hominid line and the selection pressures and cognitive developments that unfolded through its six million years of existence. And, finally, we need a sharp grasp of the architecture of the emotion-processing mind, the phenotypical unit of selection we have chosen to represent emotional cognition and adaptation. If we can put them all together into a well-synthesized accounting, we should emerge with insights previously unfathomed.

The adaptationist programme that I develop here has, then, three essential requisites:

1. A clear definition of the unit of selection, the emotion-processing mind.
2. A full picture of the fitness environment and selection pressures that fashioned the hominid mind.
3. The use of basic principles of evolutionary change as a means of accounting for the present structure and adaptive functions of the cognitive mental module.

Let us now succinctly define each of these requisites.

The essential attributes of the emotion-processing mind

In the previous chapter, I compiled a number of features of the emotion-processing mind and its two main systems—the conscious and deep unconscious systems. Here, I will distil and clarify the main properties of this module of the mind to help us to focus on the vital attributes and mysteries related to the unit of selection whose evolution we are exploring. The following seem to be the most pertinent:

1. Emotionally relevant knowledge of the world, and the behavioural adaptations constructed on that basis, derive from two fundamental sources:
    a. affective attunement and responsiveness;
    b. cognitive attunement and responsiveness.
2. The emotion-processing mind is the module or aggregate of mental capacities that is the basis for cognitive adaptations to emotionally charged environmental impingements.
3. This adaptive module is comprised of two essentially discontinuous parallel processing systems, *the conscious system* and *the deep unconscious system.*
4. The conscious system receives emotionally laden inputs through the five senses, mainly audition and vision, in a manner that involves an initial phase of conscious registration and experience. The system then brings its capacities for learning, thinking, recalling, memory storage, and the like to bear on the incoming trigger events and their manifest and implied meanings, developing adaptive responses accordingly.
    a. The conscious system contains a superficial unconscious subsystem with memory storage and unconscious contents that are accessible whole-cloth, directly (undisguised) or with minimal disguise, to awareness. There is, however, a defensive alignment that creates a conscious-system repressive barrier that safeguards consciousness from recovering selected, potentially accessible but overdisturbing contents and their meanings; psychodynamic issues play a role in this type of direct memory access. This repressive barrier is, however, very different from the barrier that keeps deep unconscious system contents from awareness—the emotion-processing mind operates with at least two distinctive forms of repression.
    b. The conscious system is essentially the organ of psychological and emotional human adaptation as it bears on

the issues of long- and short-term survival and eventual reproductive fitness. These concerns organize and give meaning to its adaptive capacities and functioning.

c. With respect to rules, frames, and boundaries, the conscious system is relatively frame-insensitive and inclined towards frame modifications.

5. The deep unconscious system receives inputs from the five senses, mainly audition and vision, but the inputs and their meanings do not register in awareness; *subliminal or unconscious perception is its mode of receptivity.*

    a. This system has two main subsystems:

        i. a *wisdom subsystem*, which cognitively processes incoming stimuli or triggers, and their unconsciously registered direct and implied meanings, and arrives at adaptive solutions to the issues posed by these unconsciously perceived environmental impingements: these adaptive responses are then encoded into the narratives communicated by patients in psychotherapy (and more generally, in all narrative communications inside or outside of the treatment situation);

        ii. a *fear–guilt subsystem*, which houses all forms of personal and borrowed guilt and the consistent fear of personal death.

    b. The deep unconscious wisdom subsystem is able to reveal the nature of its adaptive operations solely through trigger-related, encoded narratives. However, the insights and adaptive recommendations of this highly intelligent and efficient system do not impact directly on awareness; they therefore have no palpable effects on an individual's direct coping efforts.

    c. With respect to rules, frames, and boundaries, the deep unconscious system is exceedingly frame-sensitive and inclined towards secured frames.

    d. The deep unconscious fear–guilt subsystem, silently and without conscious realization, has a profound and powerful effect on direct, conscious coping efforts. In general, this influence unconsciously presses conscious adapta-

tions towards highly defensive and self-hurtful choices and responses.
4. The conscious and deep unconscious adaptive systems of the emotion-processing mind are essentially *language-based* processing systems. This proposition implies that the evolutionary history of the emotion-processing mind will be closely linked to the development and history of cognitive functions and language acquisition (Bickerton, 1990, 1995; Chomsky, 1980, 1988; Corballis, 1991; Donald, 1991; Liberman, 1991; Pinker, 1994; Pinker & Bloom, 1990).

Some unsolved puzzles

We have in hand a broad picture of the organ of adaptation that is the subject of our evolutionary investigations. It will help to orient us further if we turn now to listing some of the more unexpected and puzzling features of this mental module. These features of the emotion-processing mind are insufficiently explained by studies and postulates developed on the basis of the non-evolutionary sub-theories in the hierarchy of psychoanalytic theory, including the communicative approach, and it is our hope that our adaptationist programme will help us to account for their present existence.

What, then, are the most striking anomalous features of the emotion-processing mind? To answer, they appear to be:

1. The existence of two essentially discontinuous processing systems for emotionally charged impingements.
    a. The world of experience for each system of the emotion-processing mind is quite different. The conscious system is a widely scanning system, while the deep unconscious system is narrowly focused on frame impingements.
    b. The adaptive choices made by each system are strikingly different.
    c. There are but two means of contact between the conscious and deep unconscious systems:

i. The flow of deep unconscious system experiences and adaptive solutions from its wisdom subsystem to the conscious system via *encoded* narratives. This flow is subject to a defensive gradient that bars unencoded images from conscious awareness and resists the emergence of emotionally charged disguised themes as well. There is also an additional conscious-system defence directed against linking the encoded themes that do reach awareness with the trigger events that have evoked them—an effort that would reveal the most cogent meanings of emotionally charged events.

ii. The second means of contact is via the flow of the unarticulated, unconscious effects of the deep unconscious fear–guilt subsystem on the conscious system and its adaptations.

d. It is noteworthy that the conscious system does not influence deep unconscious processes and the deep unconscious wisdom subsystem does not affect conscious adaptive choices. The latter arrangement entails a great loss of adaptive knowledge, a loss that is further extended through conscious-system defensiveness.

e. On the other hand, the mental circuitry that does have strong unconscious links to, and unconscious effects on, the conscious system and direct adaptation arises from the deep unconscious fear–guilt subsystem. This renders the conscious mind and adaptive preferences highly vulnerable to self-hurtful actions and choices.

2. The strikingly defensive structure of the conscious system, which significantly reduces its knowledge of and contact with the environment. The result is a significant sacrifice of the critical information and meaning needed for sound adaptive choices and responses. This arrangement also provides the soil for neurosis based on repressed, unconscious perceptions, motives, and needs, and dysfunctional reactions to a wide range of unconscious experiences.

a. Despite the enormous power that deep unconscious experience and needs have over our emotional lives, the conscious system is designed with a series of massive defences designed to render it oblivious to the meanings of these experiences.

3. The propensity of the conscious system to seek out and favour frame deviations in dealing with emotional impingements—a tendency that is generally harmful for all concerned. Concurrently, this system is strongly opposed to secured frames despite their many health-giving features.

4. In contrast, the deep unconscious system not only appreciates and favours frame-securing efforts, but, as noted, is also designed to be almost entirely focused on ground-rule and boundary phenomena and issues. This means that the human appreciation for health-giving secured rules, frames, and boundaries exists largely outside of awareness where it does not affect adaptive frame-related behaviours and choices.

5. The emotion-processing mind is split with regards to frame-related interventions and moments. It reacts unconsciously to frame deviations in terms of the great harm they cause, even as they are welcomed consciously. But, on the other hand, it reacts to frame-securing moments consciously as if they will lead to personal annihilation, even as they are welcomed unconsciously. This implies that whichever direction a frame-related experience goes—towards frame-securement or frame-alteration—the emotion-processing mind is split and experiences on one or the other level a sense of danger and threat.

This is quite an array of puzzling and paradoxical features. Taken together, it would seem that the design of the emotion-processing mind is not only fundamentally flawed, but virtually dysfunctional in its adaptive operations. It would appear, then, that psychotherapy essentially can be defined as the attempt to overcome the consequences of the evolved, seemingly dysfunctional universal design features of the emotion-processing mind—small wonder that many people say that everyone needs to be in psychotherapy at some point in their lives.

The idea of an evolved organ with built-in pathological components is not entirely unprecedented. Under natural conditions, carious teeth are the rule rather than the exception, and if a man lives long enough, he is almost certain to develop prostate cancer; the knees and eyes of most people give out after forty years or so. But the emotion-processing mind is compromised from its beginnings, and it plays a pivotal role in our lives from infancy onward, so the consequences of its design flaws are pervasive and extremely serious. To be sure, we are dealing with a most serious situation.

The key principles of evolution

Having highlighted the most perplexing features of our unit of selection, let us now recount and further hone the guiding principles of evolutionary change—evolution as descent with modification. While there are several ways to describe the essential principles of evolution, the simplest yet most complete version appears to shape up as follows (see Plotkin, 1994):

1. The presence of random variations in genes and their phenotypes (the expression of genes in physical and mental capabilities)—*the generative or interactor phase*.
2. The action of natural selection on the phenotypes over long epochs of time—*the test phase*.
3. The differential reproduction of the most successful phenotypes or adaptations—the *lineage or regenerative phase*—which depends on the existence of reliable replicators such as genes.
4. Adaptations and survival strategies that are based on the operations of the selected variants. These variants are eventually challenged by changes in the environment that create the need for adjustments in existing adaptive resources and for the emergence of fresh random variants or mutations.
    a. This mixture of old and new variants is then carried forward into a new era in which natural selection operates in light of the present fitness environment and the organism's currently available adaptive resources.
5. Selection acts on phenotypes, which provide organisms with the structures, entities, and processes with which to adapt to

the existing but ever-changing environments in which they live. The genes that underlie the most favourable adaptations are preferentially replicated, thereby creating descent with favoured variation.

a. All organisms develop *survival strategies* that reflect their adaptive skills vis-à-vis their environmental impingements. The fundamental adaptive issues are those of survival and reproductive fitness that allows for genetic replication.

6. There are, however, several subsidiary considerations with regard to natural selection:

a. Adaptations are knowledge-based ways of comprehending and coping with the world and its impingements.

b. Adaptations deal with both relatively certain and relatively uncertain environmental conditions—with certain and uncertain futures.

c. Given that environments are both stable and unstable, for survival to be ensured, adaptations must themselves be stable, yet capable of a rapid change.

d. Genetically programmed adaptations—knowledge of the world—change very slowly and only after thousands of years of descent. Survival needs have dictated the evolution of additional adaptive resources, based on individual intelligence and cultural or shared intelligence—secondary and tertiary heuristics (see chapter 6).

e. To grasp the evolutionary history of the emotion-processing mind, then, we must consider not only genetically programmed adaptations, but also higher-level coping strategies based on cognitive capabilities within individuals and shared with others.

f. Natural selection is an imperfect mechanism, which, as we will see, is repeatedly faced with complex and conflictual environmental issues. As a result, trade-offs and compromises related to cost–benefit factors are the rule (Nesse & Williams, 1994).

g. There is a hierarchy of selection pressures and adaptive issues that influence adaptive choices, and their selection and replication. In addition to immediate and long-term survival and reproductive fitness, the following basic adaptive issues are of note:

   i. autonomy versus dependency—going it alone or in a group;

   ii. simplicity versus complexity—limited capacities with energy-saving features versus more costly, intricate, and more sophisticated structures and functions;

   iii. order versus chaos, regularity versus disorder;

   iv. stability versus instability, sameness versus change;

   v. instructionism versus selectionism—energy-saving enslavement to the environment versus an energically costly repertoire of inner creative resources.

  h. Natural selection involves choosing between existing variants—entities. This process implies that rather than *instruct* a passive *tabula rasa* type of organism, the environment *selects* from the repertoire of resources available within the organism.

   i. In addition to genetic make-up and environmental pressures on phenotypes, environmental factors play a role in foetal development, greatly influencing its course and the nature of the adaptive capabilities of a given organism and species.

7. Regarding natural selection, and the environmental pressures and challenges that guide and constrain this process, there is a basic set of challenges and issues that have confronted all species, including the primates and hominids. Listing these relatively universal adaptive issues serves to identify the common driving forces in evolutionary change and provide another backdrop against which we can identify distinctive adaptive issues for the emotion-processing minds of each of the hominid species we study.

Among the relatively universal selective pressures and adaptive issues that are confronted by virtually all species, the following are most relevant to our pursuit:

  a. Immediate and long-term personal survival, which is the province of conscious adaptations and essentially reflected in an organism's *survival strategies*. In part, the nature of this strategy depends on the extent to which an

organism is aware of the future, compared to being relatively locked into the present. In this connection, there are three key adaptive issues:
   i. defence against predators;
   ii. obtaining supplies of energy;
   iii. being free to reproduce.
b. To some extent all organisms must deal with issues related to the rearing of offspring.
c. Similarly, all organisms must deal with their relationship with conspecifics—same-species others.
d. All species must, to varying degrees, adapt to the issue of personal death and deal with the selection pressures created by this issue. While genes are indifferent to death (even though they are designed to be reproduced), the organisms they create are not.
   i. There is, however, a crucial evolutionary history to the awareness of self and conscious realizations of the inevitability of personal death and the loss of important others, which reaches its maximal sharpness in *Homo sapiens sapiens*.
   ii. For humans, then, concerns about death and the expectation of personal demise constitute emergent, unique, internally driven selection pressures that greatly influence both immediate adaptive choices and evolutionary history.
   iii. The role of death anxiety in evolution, especially as it applies to hominid species, is perhaps the single most neglected issue in Darwinian theory—and in psychoanalysis as well.

Some final perspectives

In approaching our evolutionary scenario, one final contextual note is needed. There has been an unfortunate tendency for evolutionary biologists and psychoanalysts to idealize the choices made by natural selection and to view virtually everything that human beings experience and enact as adaptive in some fashion (see especially, Nesse, 1990b; Williams & Nesse, 1991). While it is true that every organismic response is an *effort at adaptation*, it does not follow that the effort has been successful in arriving at an optimal coping strategy. This is especially true for the evolution of *Homo sapiens sapiens*, in which the selection pressures have been and are enormously complex, the amount of time available for the expression of variation and selection is often quite limited, and trade-offs and compromises are absolutely inevitable.

This part of the picture will be greatly clarified when we develop definitive criteria for successful adaptations (see Bock, 1980; Mayr, 1974, 1983; Plotkin, 1994; Tooby & Cosmides, 1990a, 1990b; Williams, 1966, 1985). For now, all we have to work with are crude assessments of the cost–benefit ratio of a given adaptive effort and adaptationist programmes that are constructed in relatively sound, unbiased, scientific, testable, illuminating fashion.

Simon (1962, 1982), Gould (1980, 1982, 1987, 1989), Lewontin (1970, 1983), Tooby & Cosmides, (1990b), Plotkin (1994), Wesson (1994), Eldredge (1995), and Dennett (1995), among others, have stressed the complexity of the forces that influence natural selection. There are distinct engineering, design, and other limitations with respect to the options available to natural selection in relation to the evolutionary possibilities available for the development of a given system. There also are natural restrictions to the range of potential and actual variants, to the structural possibilities that apply to a given organ system, and to the environmental pressures that affect the overall design and coping capacities of living organisms—and their possible range of evolutionary changes. Simon (1982) has used the term *satisficing* to characterize most adaptations, a term that recognizes that, in general, adaptations tend to be practical but compromised solutions to the environmental issues to which they are responsive.

We are well warned again that 99% of all of the species that have ever existed on this planet were unable to generate adaptations that would have allowed for their continued existence. Adaptive failures have far outweighed adaptive successes, and compromise seems to be the norm. Such indeed, as has already been suggested and, as we later see, appears very much to be the case with the emotion-processing mind. To understand just what this means, let us turn now to the specifics of the adaptationist programme forged by the communicative approach for this mental module.

CHAPTER TEN

An evolutionary scenario: the early hominids

My intention now is to trace the development of the emotion-processing mind through the history of the hominid species, beginning with the first species in the line, and to move forward from there. The focus in the present offering is on the phenotypical configuration that takes the form of the emotion-processing mind. I propose to concentrate on five dimensions relevant to the adaptationist programme that I intend to narrate.

Before presenting these guidelines, I want to acknowledge the good fortune of being able to draw on the work of Donald (1991) for this endeavour. His masterful presentation of the evolutionary history of language and cognitive functioning helped me to organize and find my way through the maze of anthropological investigations that were needed as a background for the adaptationist programme I offer here (see for example, Barkow et al., 1992; M. Brown, 1990; Calvin, 1991; de Duve, 1995; Fagan, 1990; Johanson & Shreeve (1989); Kuper, 1994; Leakey & Lewin, 1992; Ridley, 1985; Ward, 1994). With their help, my exposition is organized around the following rubrics:

1. The general state of the hominid line, of which there are four subspecies:
   a. *Australopithecus*. First appeared: 4–6 million years ago;* key cognitive-related event: bipedalism.
   b. *Homo habilis*. First appeared: 2 million years ago; key cognitive-related event: cranial enlargement with striking brain asymmetry.
   c. *Homo erectus*. First appeared: 1.5 million years ago; key cognitive-related event: very rapid increase in brain size.
   d. *Homo sapiens*. Archaic form, first appeared: 300,000 years ago. *Homo sapiens sapiens*, first appeared: 200,000 years ago; key cognitive event: modern vocal capabilities and auditory apparatus enabling the development of language.

Each hominid species dealt with a distinctive, broadly defined, stable yet changing environment that included such dimensions as social-cultural norms, the physical settings of adaptation, modes of communication, general cognitive skills and resources, and personal physical attributes. Nevertheless, despite occasional radical changes in these complex environmental impingements, the time periods between the emergence of new species were on the order of millions of years.

2. I explore and integrate four distinct yet interrelated processes and events for each hominid species. They are:
   a. the background situation that defined the species and its fitness environment;
   b. the basic cognitive capabilities that were available to support emotional cognition;
   c. the main selection pressures, including the specific problems of adaptation that needed to be solved;

---

*These dates are approximations, drawn mainly from Leakey and Lewin (1992) and Donald (1991). The specific timing of the appearance of these hominid forms, and the debates about their identity, exact features, and the like, need not concern us here.

d. the transformations that took place in the design of the emotion-processing mind, their cost–benefit ratios, and the nature of evident competencies, compromises, and inefficiencies.

The following, then, is an evolutionary adaptationist programme fashioned to account for the development of the emotion-processing mind.

The Australopithecines

Background situation

The first hominids split off from the primates six million years ago and were distinctly bipedal, with erect posture, freed hands, and appositional thumbs. They showed a striking increase in brain size compared to the apes, and they were strongly visual. They survived mainly through their generally increased intelligence and ability to see over long distances. Building on faculties seen in their predecessors and compared to the apes, *Australopithecus* showed advances in the following areas: socialization, social stability including pair bonding and family structure, infant care carried out by both parents, food sharing, and the use of home bases.

General cognitive capabilities

While the Australopithecines were far more social than the apes, they showed few cognitive advances and little cognitive re-structuring. They shared with the apes such cognitive capabilities as curiosity, imitation, deceit and revenge, focused attention, memory, reasoning and problem-solving, and primitive, pre-linguistic forms of cognition.

The main form of cognition in *Australopithecus* was *episodic*, in that it was concrete, situation-bound, and non-reflective— time-bound to the present moment. Learning was procedural and involved actions rather than thought, although there were indications of a crude capacity to develop internal representations and of abilities to abstract from concrete events. Social learning was in evidence, but it, too, was based on event perceptions and concrete instances, without a capacity to represent a situation mentally and reflect on it in its absence.

In all, then, the Australopithecines possessed a wide array of cognitive abilities, which were, however, bound to immediate moments and concrete situations. Conscious intelligence was pre-language in form and action-bound. There were signs of conscious emotional issues and of their processing in the immediate moment. However, there is no evidence of an emotionally related, psychodynamic cognitive *unconscious* system of the mind, although there were capabilities for automatic activities for which awareness did not intercede.

Selection pressures, adaptive issues

The Australopithecines were faced with several emergent adaptive issues and selection pressures beyond those faced by the primates. They included a dawning social organization of greater stability and complexity compared to that seen with apes. Furthermore, because the species' brain had increased in size, the design dynamics of feasible female pelvic size versus tolerable foetal cranial size for birth passage led natural selection to favour offspring who were born with smaller brains than their predecessors. This in turn increased brain immaturity and seriously limited the post-natal survival skills of offspring. This trade-off therefore necessitated an extended period of weaning and parental care to allow for brain and mind development to the point at which the child could survive relatively independently.

Social organization and sociocultural factors emerged as especially strong adaptive challenges and selection pressures. Over the remarkable time period of some two-and-a-half million years, these selection pressures led to increasingly complex social structures and relationships, but to little in the way of basic cognitive change. For example, *Australopithecus* did not show a capability to make and use tools; had they done so, it would have suggested a break-through in cognitive development. A long period of stasis, perhaps some two million years, set in during which natural selection did little to change this first hominid species cognitively and otherwise.

Transformations in the design of the emotion-processing mind

What, then, was the architecture of the emotion-processing mind in the Australopithecines? They did, of course, possess a conscious system, which enabled them to negotiate directly the emotionally charged issues that arose in their expanding social interactions. While it is likely that there was an incest taboo in place—it exists for apes and other vertebrates (Brown, 1991)—there may also have been some type of dominance hierarchy. Problems in controlling violence are likely to have been the primary intra-species and inter-species adaptive concern. Combative cognitive skills would then have been favoured by natural selection.

As for the mental processing of emotional impingements, event perception and the confinement of processing efforts to immediate situations evidently limited this species to *conscious-system processing*, with some assistance from non-language memories of prior similar events, activated only when they occurred again in some form. Most of these adaptive capabilities were instinctive or genetically determined, although some minimal qualities of individual intelligence may also have played a role. Shared or cultural wisdom was probably absent or minimal in extent.

There is no research that supports the presence of definable *unconscious perceptions* in apes, and it probably did not exist in the Australopithecines. This type of perception, which may draw on precursors in abilities related to peripheral vision, may well require a language-based, automatic sorting capability in order to establish two separate perceptive and processing systems—one accompanied by awareness, the other operating systematically, but without awareness. Meaning detection requires abstracting, conceptualizing, and representing, as well as thoughtful processing. None of these abilities were within the grasp of the Australopithecines in anything but the crudest of forms.

All in all, it would appear that the emotion-processing minds of the Australopithecines were single-system minds, conscious processors with striking limitations because of the lack of language capabilities and the inability to move beyond the immediate moment. Unable to represent the world abstractly in the absence of a given set of events, superficial unconscious memory storage was greatly constrained. Recall was strictly prompted by an immediate event that conjured up an earlier incident.

There is no suggestion of repression or denial in *Australopithecus*, although repression may well have existed in a crude form as some kind of automatic exclusion mechanism directed against painful memories when fresh, relevant events linked to past traumas took place. It is commonly accepted that nature seldom creates a mechanism *de novo*; it tends to use existing designs and tendencies and to convert them for new uses. With regard to repression, then, the communicative view is that the mechanism is a broad-based, fundamental, defensive mental mechanism designed to protect self and others. But repression appears to depend strongly on internal representations based on language abilities and can operate only in the simplest manner possible without language.

As for denial, as we will see, its development and use is largely a consequence of advanced forms of self-awareness, which, in turn, lead to the awareness that personal death inevitably lies ahead. Capabilities of this kind were lacking in the Australopithecines, and the selection pressures created by this issue evidently did not exist. Thus, there is little reason to believe that *Australopithecus* used denial in any form. At their level of development, denial would involve the obliteration of danger situations, with the potential for harm and death. Denial is a mechanism that dynamically reduces an organism's awareness of his or her environment, including life-threatening predators. It therefore seems unlikely that the mechanism would be favoured by natural selection in the absence of death anxiety, because its use diminishes chances for personal survival.

In sum, then, it seems likely that the emotion-processing mind of the Australopithecines was a single-system mind, with conscious processing capabilities and a superficial unconscious memory system that had little psychodynamic status and little in the way of psychological defensiveness—either repressive barriers or denial mechanisms.

Homo habilis

Despite another one million years of developmental time, this species of hominids was very much like the Australopithecines—attesting again to the incredibly slow pace of evolutionary change. While the brain continued to grow in size, socially and cognitively *Homo habilis* seems to have been little different from its predecessor, the Australopithecine. The main distinction for *Homo habilis* was the development of the first stone tools, the dawning of a soon-to-unfold cognitive revolution. The emotion-processing mind of this species, then, was probably much like that of the Australopithecines.

Homo erectus

Background situation

This species was pivotal in the transformation from ape-like features to those that are human. Compared to its predecessors, *Homo erectus* showed a dramatic increase in the making and use of tools; they hunted cooperatively, created fires, cooked their food and were the first to leave evidence of consistent meat eating, developed relatively stable home bases, and had an expanded social structure with capabilities for the cultural transmission of information, including pedagogical teachings.

General cognitive capacities

With *Homo erectus*, there was a shift in culture and communication from being entirely episodic to a culture and mode of expression that was largely *mimetic*. Thus, cognition remained without language, but advanced to the use of gestures and mimesis. Cognition was essentially visual with categorical perceptions, generative patterns of action, and comprehension of social relationships. An important aspect of mimetic skill is the ability to initiate representational acts consciously, an indication of an ability to sustain clear internal representations of people, places, and events.

The key properties of the mimetic mode of expression and culture are intentionality, a diminution of egocentricity with a growing awareness of others and their needs, generativity, social communication, the use of referents that are distinguished from events themselves, having unlimited objects available for representation, and autocueing—the use of self-generated cues based on non-language representational thinking. While mimesis is concrete and an episode-bound medium of expression, it is able nevertheless to represent an unlimited number of episodes.

There were many social consequences to the development of mimesis, including a great increase in the complexity of social structure and interactions. There was reciprocal mimesis, games, controls and customs, clearer individual role definition, the sharing of knowledge, organized group activities, primitive rituals, the beginnings of innovation, a modest increase in the rate of cultural evolution, group migration, stronger pair-bonding and a more elaborate and extended family structure than with the previous hominid species, an increase in the number of offspring per mother and a lengthening of the weaning period, a more notable division of labour among group members, increased pedagogy, and greater relative stability to home bases with fixed shelters and yet more migration as well.

Essentially, the cognitive basis for mime is:

1. The ability to represent one's self and body.
2. An increase in conscious motor-control capacities—mime is a deliberate action form of expression.
3. An increase in the sense of the existence of others (a diminution of egocentricity) and a greater ability for event perception.

All in all, *Homo erectus* was a species that showed a rapid increase in brain size and an explosion of fresh cognitive capabilities based on mimetic expression and communication and on a dramatic increase in the complexity of social relationships, interactions, and structure. The ability to develop and manipulate internal representations of events produced non-language thinking and reasoning, yet, in over a million years, language itself did not emerge.

Selection pressures

The main intensification of adaptive challenge occurred in the social-cultural realm. There were increased family demands; highly complex forms of social cooperation, orderings, and relationship issues; and a greater complexity to life in general. While there were early signs of simple rituals, there was no indication of a definitive awareness of the future and no sign of an extended sense of personal identity that would define one's life span. Those who died were not buried, though they were mourned.

Adverse physical environments were challenged rather than avoided by *Homo erectus*, and protection against predators and the means of gaining food resources—survival strategies—became more sophisticated and took on a well-defined cooperative–social structure. Energy sources expanded into cooked meat, while vegetational sources still relied on what nature provided—growing one's own food was still hundreds of thousands of years in the future.

Little seems to be known about intra-species violence, but there is evidence of warfare, conspecific killings, and violence across as well as within groups and emerging clans (Alexander, 1989). As tools were refined, weaponry became a factor, making the control of violence all the more a major issue. Overall, then, selection pressures seem to have centred on negotiating an increasingly complex social life and in developing the means of surviving predators and conspecific attacks.

Transformations in the design of the emotion-processing mind

Conscious-system capabilities in *Homo erectus* clearly had expanded greatly from those of its predecessors, allowing for a greater degree of intelligence for use in general adaptation, survival, and reproduction. Improved hominid intelligence was absolutely essential for the survival of *Homo erectus* in the savannah and in the many new environments they entered—jungles, forests, and lands overrun with predatory animals.

Man and woman living and surviving by wit alone is a close approximation. But to do so, the function of consciousness needed to be honed, with a stress on increasing the sharpness of sensory perceptions and by developing a highly focused and concentrated form of conscious attention. Variants that possessed these particular mental features were evidently selected for differential reproduction, a fateful turn for the evolving design of the emotion-processing mind.

Perhaps most critical in this regard was the development of focused attention so that only small segments of information could register in awareness at any given moment. The conscious mind evolved to carry out the serial processing of information and meaning, one bit at a time, and this type of processing was fashioned as a very sensitive and delicate adaptive capacity that was highly concentrated but easily disturbed.

One can imagine a form of awareness capable of registering multiple simultaneous inputs (see chapter 14). However, survival evidently was not served by such a capability, and it was selected out and became extinct—if it ever existed. On the other hand, the limitations to the amount of information and meaning that could be processed by the conscious system were set in place—for example, the well-known capacity for the conscious mind to process 5 to 9 bits of cognitive information per second (Miller, 1956). This provided *Homo erectus* with a sharp sense of what was happening in their environments and an ability to focus down on the reality events with which they were being confronted. But this design feature considerably restricted the conscious system's processing capabilities—there was an ability to generate a rapid response to danger, but with greatly restricted amounts of available information.

Another likely prominent feature of the conscious system of *Homo erectus* was its above-noted vulnerability to distraction. Concentrated attention and non-language forms of thought were the essential basis for their survival strategies, especially at moments of acute threat. The conscious system therefore developed an intensely focused responsiveness, but this also meant that distractions had to be blocked out at all costs—an unprotected attention system could easily be diverted at moments critical to immediate survival. In selecting for concentration, then, vulnerability to distraction became a critical issue, and mechanisms to reduce diverting stimuli needed to be put into place.

It seems likely, too, that the conscious system began to be overburdened and to move towards developing a critical problem of *information-meaning overload*. There was a rapidly mounting growth in the complexity of the environment, largely in the form of a changing and widening sociocultural scene; the development of complicated, multiple social and work roles and relationships; and greater risks taken in hunting, migrating, and other venturesome endeavours. These selection pressures evidently moved the mind towards the development and enhancement of mechanisms that allowed for rapidly responsive automatic actions for which awareness was by-passed. This was one of the precursors of the deep unconscious system that was soon (in terms of evolutionary time) to evolve.

What, then, of the enormous increase in emotional load—of *emotionally charged adaptive issues*? What design features of the emotion-processing mind were developed to handle, adapt to, and survive the avalanche of emotional impingements that fell upon *Homo erectus*?

To answer, it seems likely that the development of more elaborate internal representational and processing capabilities allowed for greater sophistication in the conscious system's adaptations to social impingements. Visual perception and motoric action were the dominant modes of experience, expression, and adaptation. Given the easy distractibility of the conscious system as it went about its survival functions, it seems fair to postulate that system overload soon became a critical survival problem.

Natural selection's way of dealing with informational overload leans towards creating additional systems to handle the excess (de Duve, 1995). A prototypical event of this kind occurred early in the evolutionary history of biological organisms when RNA, the first replicator, became overloaded with the information needed for cell formation and replication. The result was the selection for a second system, DNA, which then became the prime replicator and the key to the evolution of individual entities.

Given the expanding capabilities and representational abilities of the *Homo erectus* mind, a primitive form of perception without awareness may have begun to develop—a means of registering incoming stimuli without their impinging on consciousness. This development would have freed the conscious system from having to deal with an excessively large quota of information and meaning and enhanced the system's immediate adaptive, survival-related functioning. It is difficult to be certain, but at first these unconscious perceptions may have been absorbed by the developing deep unconscious mind and not represented in awareness directly or with disguise. A more likely possibility is that these unconsciously received impingements were crudely processed without awareness and then encoded and conveyed in an individual's mimetic communications.

In principle, then, the danger of conscious-system overload appears to have created a survival-threatening adaptive issue to which evolution—in the form of random variation→testing and natural selection→preferential reproduction—was compelled to respond. The most likely scenario is that there then followed the development of a mechansim through which the conscious registration of many distracting emotionally charged impingements was by-passed. This new mental function drew upon already existing capacities for automatic thinking and coping without awareness and made use of representational abilities that allowed for mimetic-based, internal unconscious processing, however primitive and crude.

At best, these developments were rudimentary, even though they were essential for survival. They probably began to unfold in *Homo erectus* because of the great increase in emotionally charged impingements that assaulted their conscious awareness and taxed their minds' attention capacities. If these multiple emotionally charged inputs had not been reduced in frequency and intensity, the conscious system would have malfunctioned repeatedly, and chances of survival would have been severely reduced.

Finally, with respect to the fear–guilt subsystem of the deep unconscious system, it seems likely that crude rudiments of guilt and fear were experienced by *Homo erectus* in some kind of action-mime, non-language form—consciously rather than unconsciously. A more fully developed deep unconscious fear–guilt subsystem awaited further evolutionary developments, especially language acquisition, and the accompanying awareness of personal mortality and of destructive impulses and acts against others and self. These developments announced the emergence of *Homo sapiens sapiens*, the hominid species to which we now turn.

CHAPTER ELEVEN

Another scenario: *Homo sapiens sapiens*

A*rchaic Homo sapiens* was a transitional species that formed a bridge from *Homo erectus* to modern-day *Homo sapiens sapiens*, our own species, which has been the exclusive hominid form for the past 150,000 years or so. I will, however, for ease of presentation, confine my remarks to *Homo sapiens sapiens* and focus on the critical developments of the emotion-processing mind that have taken place since our species appeared on the evolutionary scene.

Background situation

Homo sapiens sapiens experienced a huge increase in brain size, with strong brain lateralization. This allowed for the left-sided concentration of the brain centres needed to develop the speech apparatus necessary for language expression and the complementary auditory receptivity. There were also enhanced motor skills, changes in the vocal chambers essential for distinctive phonation, and a mushrooming of social–cultural developments that once again considerably increased the complexity of life in general and of emotional life in particular.

While typically these developments were slow in the making, they also unfolded far more rapidly than earlier evolutionary changes. Speech existed for some 100,000 years before its effective use for internal representations and communication; culture expanded, but agriculture is only 10,000 years old; and with this new-found ability to grow our own food, there was the dramatic emergence of well-populated settlements that led to urbanization and the expansion of culture far beyond anything previously known. Writing and recorded information and knowledge are only 6,000 years old, and recently there has been an explosion of information and meaning to the point where external storage in books, films, computers, and the like has become both a new form of memory and a vital necessity (Donald, 1991). And with all of this—especially the increasing complexities of external and internal emotional life, much of it quite raw and impulse-ridden—an overwhelming collection of new adaptive issues confronted this latest species of the hominid line.

General cognitive capacities

The main evolutionary development for *Homo sapiens sapiens* was, of course, the use of language for both internal representation and as a means of high-speed communication and adaptive processing (Liberman, 1991). Other advances in cognitive capabilities entailed enhanced motor control and speech abilities, voluntary control of vocalization, language and semiotic skills that included the capacity to create and manipulate symbols for events and ideas intentionally, an expansion of the use of rules and kinship regulations, and an entirely new survival strategy.

Additional achievements that belong on this large list of fresh developments are the domestication of fire and, later on, of animals; strong capabilities for problem-solving with marked increases in shared knowledge and collective problem-solving and adapting; new protection strategies against predators; and the discovery and use of vitally new sources of energy. There were also new forms of information processing and storage, and fresh capacities of explanation, understanding of causality, prediction, and enhanced control and manipulation of one's own environment with remarkable abilities to create one's own surroundings and world of experience—including aspects of one's own evolutionary selection pressures and history as well.

Finally, there also were new faculties that provided an ability to generalize across events and to extract themes across incidents. Conceptualizing and abstracting advanced greatly, and reasoning, thinking, and adapting markedly improved, including a new form of representational intelligence—the use of mind tools and mental models.

There were many consequences to these remarkable advances, including another burst of complexity in social roles and relatedness, the eventual shift to highly complicated urban living for many members of the species, and an enormous amount of cultural expansion on all fronts. But in addition to these rather familiar developments, we should stress certain cognitive developments that arose largely from language acquisition, because they played a significant role in the evolution of the emotion-processing mind (Bickerton, 1995). They are:

1. Language development promoted a highly distinctive awareness of self and of others—an increase in both ego centricity and decentricity.
2. Language and its new representational capacities not only allowed the mind to grasp and sustain realizations related to past events, and to represent them internally, but also allowed for a distinctive grasp of future events as well.
3. With this new sense of identity and sharpened grasp of past, present, and future, there emerged an early-in-life awareness of the inevitability of personal death and the death of others. *Death anxiety* became a significant cognitive factor—and an adaptive and selection issue as well.
4. Language and awareness of personal mortality brought with them the emergence of burial practices, rituals, and symbols related to the death experience, along with the origins of religions and the appearance of seers, shamans, priests, and other religious leaders.

5. These new capabilities also led to the creation of bodily decorations, clothing, and sophisticated tools and to a development related to death issues—namely, the invention of more and more lethal weapons.
6. Issues of aggression and violence escalated, especially with regard to same-species violence, in part because of battles for energy supplies, but more deeply on an emotional level, as a reaction to the helplessness caused by the awareness of personal mortality. Among other indications of unbridled conspecific violence, cannibalism was in evidence. Intra-species and inter-species competition greatly intensified.
7. Another crucial development was the capacity for story-telling, the creation of narratives, a faculty seen with mime and greatly expanded as soon as language emerged. Shared anxieties and adaptive issues were dealt with by collective story-telling and myths of origin, death, warfare, kinship, and much more. Narrative ways of conscious coping became highly developed, and the culture of *Homo sapiens sapiens* was at first a *mythic culture*. Inventing symbols for representation, thought, processing, and communication is a distinctly human capability.
8. In time, two modes of thought came into existence (see also Bruner, 1990): the narrative mode and the paradigmatic or scientific mode, each serving different functions. That is, in general, narrative thought served imagination and ultimately, as we will see, emotional cognition and the regulation of behaviour, while paradigmatic thought served science and our search for logical truths and led to the emergence of a *theoretic culture*.
9. Language increased the capacity for self-awareness and self-observation, which is, however, still quite limited and rudimentary in its range and functioning (Langs, 1993a; Ornstein, 1991). There also emerged an ability to reflect on one's own reflections and the development of distinctive and elaborate memory systems.
10. Emotionally, language gave form to an inner mental life that was both inventive and generative, and yet raw, with feelings such as violence and hate, readiness for retaliation, and

unbridled sexual urges, to a degree that was quite out of keeping with the emerging sense of civility and cooperation that was called for through cultural mores and standards.
11. Finally, there was the emergence of the ability to speculate about reality, the use of logic, and the development of science, invention, external memory storage, and information-processing systems—developments that will carry us into the future and greatly affect evolutionary changes yet to come.

Selection pressures: adaptive issues

As this catalogue of advances indicates, through natural selection, *Homo sapiens sapiens* had been afforded an extraordinary array of emergent and gifted adaptive capabilities, but the species also had to deal with an escalation of threatening adaptive problems far beyond anything faced by its predecessors. And in addition to all of the earlier and now expanded selection pressures, a relatively new type of pressure emerged—demands that stemmed from the human capacity to think and to grasp the past, the present, and especially the future.

Thus, a major selection force developed from the capacity for symbolic–representational language to portray future dangers, particularly the inevitability of personal death. The driving power of external environmental impingements as a factor in evolution was supplemented in *Homo sapiens sapiens* by the driving power of inner emotionally traumatic realizations. Among these internally derived impingements, death anxiety, including the fear of annihilation by others and the inevitability of personal death, played a far greater role than inner instinctual drives, whose expression was partly genetically determined, partly learned, and partly a component of the human repertoire of adaptive resources vis-à-vis environmental forces.

Issues of death and violence have become major aspects of the selection pressures that have and will shape the evolution of the hominid line and of the emotion-processing mind for thousands of years to come. We alone are compelled to adapt to the anticipated loss of loved ones, friends, enemies, and our own lives, and to the knowledge of the possibility that we may even destroy ourselves as a species—that is, cause our own extinction, much as we have already for thousands of other species (Leakey & Lewin, 1992; Ward, 1994). The adaptive issues raised by the grim awareness of personal mortality and the proliferation of human resources generated for the destruction of conspecifics loom large. They are a significant part of the realization that, for better or worse, humans themselves have become a major source of evolutionary pressure and change.

Transformations in the design of the emotion-processing mind

Before defining the fresh advances in the design of the emotion-processing mind found with *Homo sapiens sapiens*, let us summarize the main new adaptive issues and evolutionary pressures confronting our species:

1. A striking increase in social-cultural stresses and impingements, a complexity of life far beyond anything experienced before. Emotional life in particular grew in importance and richness—and conflict—in ways not seen earlier; internalized conflict became a factor in emotional life, and external conflict intensified to unprecedented levels.
2. The awareness of personal mortality from early childhood on.
3. Although protection from other-species predators was relatively secured, surviving conspecific violence became a major adaptive issue. The tremendous lethal power of evolved weaponry was a key factor in this development.
4. Finally, we may note the expansion of well-defined stresses that stemmed from language acquisition and language-based communication—our most gifted adaptive tool brought with it a huge load of adaptive problems.

It is essential to appreciate the enormity of the increase in emotionally charged impingements faced by *Homo sapiens sapiens* as compared to the apes and early hominids—our species can also be said to have created an *emotional culture*. Relationships, social and work roles, social structure, and socially related dangers were extremely important to this species, and they brought with them powerful, unrelenting emotional impingements and adaptive challenges.

For example, an infant's need for long-term nurturing well into maturity was accompanied by the inevitability of parent–child and parent–parent conflict, and this led to strongly violent impulses and fantasies within the mothers and fathers of these infants, and within the infants themselves. But there was still more: the parents' responsibilities for child rearing also conflicted with their other roles and assignments and aspects of their needs for self-fulfillment, especially for the mothers—and further discordant and emotionally disturbing interactions between parent and child ensued. In addition to a great increase in emotional trauma, virtually everything that was gratifying emotionally brought with it a measure of conflict, pain, and the potential for emotional turmoil. For the developing child, whose mind was just beginning to forge effective adaptive capacities, these assaultive inputs and frustrations from individuals on whom the child's life depended created unprecedented adaptive issues that would be difficult to resolve through natural selection in the little evolutionary time that our species has existed.

Homo sapiens sapiens needed to forge new resources in order to cope with the enormous selection pressures that it had created for itself, but the time constraints of genetically driven evolutionary processes turned this situation into a biological emergency that nature was quite unprepared to deal with. The relatively brief evolutionary history of the emotion-processing mind is a major factor in its limitations and compromised selective choices.

The intensity of these adaptive issues also reminds us that our evident improvements in intelligence, inventiveness, mental representations of the world, quality of life, and such emerged through natural selection with a large price-tag attached to them. The concept of an idyllic outcome to natural selection and the idea that an effort at adaptation is, per se, non-pathological are essentially untenable. We see again that natural selection operates mechanically in an atmosphere of conflicting selection pressures that make compromise and imperfections—and even failed outcomes—inevitable.

Key selection pressures

We are moving towards articulating the likely scenario played out as natural selection chose from random variations in the structure and design of the emotion-processing mind in light of its goal of selecting for effective adaptations to this staggering excess of emotionally charged impingements. The following appear to be the major issues that natural selection has had to negotiate in this regard:

1. Emotionally laden impingements were so pervasive, varied, intense, conflictual, so often directly or by implication matters of life and death, dangerous, at times incestuous but more often involving violence, and so complex and demanding of attention and coping responses that they threatened to *overload* greatly the conscious system's processing capacities. Were it not for the development of the deep unconscious system, this situation would have been disastrous.

2. Language acquisition added intensity and meaning to a wide variety of disturbing perceptions, primarily of others and secondarily of oneself, that had to do with wishes, impulses, and tendencies that boded enormous hurt and injury, physical and psychological, to the perceiver and others. As human animals, we are filled with impulses of revenge, rivalry, jealously, envy, and frustration. These emotions generate a wide range of behaviours and communications that directly and indirectly convey these dangerous inclinations. These range from outright attempts at murder to lesser but terrifying substitutes for such intentions, including verbal–affective expressions that are conveyed manifestly or latently through multi-layered physical and language-based communications.

3. The amount of disruptive emotionally charged meaning and information being communicated from one individual to another, and being experienced internally within a given person, threatened to disrupt the adaptive capacities of the conscious mind. A state of *conscious-system overload* either materialized temporarily or was threatened, and without the development of mental and/or physical mechanisms to reduce the level of emotional bombardment on the conscious system, chances of survival would be significantly dimin-

ished. *Homo sapiens sapiens* may well have experienced a crisis of nature in which the survival of the hominid species was at stake. Indeed, it may well be that the failure of the Neanderthals to solve this adaptive problem, largely because of their absent or minimal language capabilities, contributed to their eradication as a subspecies.

4. To this intensification of emotionally charged impingements was added the terrifying prospect of personal death and the eventual or more immediate loss of loved ones and others. Some means of dampening these morbid realizations in order to reduce greatly their disruptive effects on the conscious mind was sorely needed.

Some possible designs for the emotion-processing mind

Let us try to imagine the possible mental variants that could have emerged, mainly by chance, for natural selection to test and choose from. What are the conceivable strategies and designs of the emotion-processing mind that could have been forged to deal with these overwhelming issues?

To answer this exercise in genetic engineering (Dennett, 1995), I first offer some possibilities that apparently have not as yet materialized. Doing so will give us a sense of the difficulties natural selection faced in choosing among solutions for dealing with the conflicting forces with which the emotion-processing mind was and is beset.

1. *The defensive obliteration solution.* This is a possible design in which a large number or all of the potentially disruptive emotionally charged impingements simply would not register consciously. This could be accomplished by selecting for a set of mechanisms that would render the conscious mind impervious to psychological threats, while retaining a sensitivity to physical danger.

   Basically, this choice would require the operation of both perceptual and psychological defences, such as denial and repression, capable of acting rapidly, automatically, and effectively to screen out most emotionally disturbing impingements fully. But these psychological defences would need to be so extremely obliterative that there would be no adaptive processing of these inputs because they would not obtain mental registration and representation. We might term this *the complete ignorance solution*.

In terms of cost–benefit ratios, the benefit side of this adaptive choice would be the considerable protection of the conscious system from disruption due to emotional issues—chances of survival thereby would be increased. But the cost would entail the elimination of the conscious realization of many psychologically harmful experiences and of many warning signs that could turn out to be preludes to later attempts at physical and/or psychological harm against the perceiver. This design of the emotion-processing mind would, then, render the individual quite insensitive to conspecific predators and vulnerable to annihilation. In addition, perceptions of others and self would be flat and without emotional valence, creating relationships that would be mechanical, dull, insensitive, without due affect, unguarded, emotionally damaging, often inexplicable, and uni-dimensional. The costs here seem far to outweigh the benefits.

2. *The desensitization plan.* This choice would involve adjustments in the conscious mind that would enable individuals to be aware of the flood of socially traumatic inputs, but would also enable them to be impervious to the meanings involved. This option would also preclude the adaptive working-over of threatening emotional stimuli, and the insensitivities involved could be highly detrimental to all concerned. This selection could be described as the *rigid emotional armour* solution.

Here, too, conscious functioning would be safeguarded, but at the cost of emotional deadness of a kind that could render individuals dysfunctional in the social sphere. It very much seems that whatever solution we conjure up, there is an inordinately high price to be paid for choosing it. This enables us to appreciate the incredible dilemma that natural selection was and still is facing in selecting for the emotion-processing mind.

Protecting the survival functions of the conscious system requires a level of defensiveness and insensitivity that comes with great cost attached to it—or so it seems.

3. *The sensitive but insensitive strategy.* This choice would allow for the conscious registration of disturbing emotionally charged inputs, but entail a conscious mind that would be aware of emotional threat, yet able to stay calm and unperturbed by it. This is a radical solution with minimal cost, but it would be entirely unprecedented historically, and it would also reduce overall adaptive effectiveness because of a lack of motivation for turning to much-needed coping responses in the face of emotional danger.

The main problem with this particular solution is that it, too, would entail an emotional insensitivity that would again be a very costly design feature. In addition, as far as I know, this would be entirely without precedent. Organisms have been selected and have evolved almost entirely because they are more sensitive than their competitors to their environments. Nature has repeatedly chosen designs of the mind that enhance knowledge of the outer world and the organism's responsiveness to specific stimuli. Choosing for an emotionally insensitive emotion-processing mind would require an entirely new design plan because it would ask humans to be insensitive knowingly to the very kind of inputs that organisms throughout evolutionary history have been sensitive to. In addition, it seems all but impossible to experience and not experience emotionally charged meaning simultaneously—the design itself, while promising, may be unfeasible engineering-wise.

The favoured choice

In all, then, the essential dilemma in designing the emotion-processing mind appears to be that in order to increase chances of survival, an organism must know as much as possible about its environment and itself. But there are aspects of the emotional environment, especially the emotionally charged intentions of others and one's own inner thoughts and impulses, that are truly unbearable to awareness, so that their reaching consciousness would disrupt *conscious-system* functioning. Knowledge acquisition, which has been the optimal adaptive choice until now, became a liability and a costly selection for *Homo sapiens sapiens*.

Imagine a mother who wishes to cannibalize her child, a father who wishes to disembowel his son, a scientist so envious of the successes of his colleagues that he wants to commit mass murder—and whatever other horrors you can conjure up. Now compound these images thousands of times over, and you can sense what's at stake. How, then, did natural selection choose a design fashioned to solve this problem, and with what benefit and cost? How perfect or imperfect is the existing architecture of the emotion-processing mind?

To answer this critical question, it will help to deconstruct the design problems facing natural selection into three separate though interrelated issues: first, dealing with the excess of input; second, creating a processing mechanism to deal with the inputs that are received unconsciously by the emotion-processing mind; and, third, solving the problem of what to do with the results of these processing efforts. The solutions to each of these challenging issues needed to be integrated smoothly into a well-functioning emotion-processing mind in a way that, overall, most favoured adaptive success. Let us look now at how natural selection opted to solve each of these problems:

1. *The input problem.* Given the mounting conscious system overload generated by both the magnitude of emotionally charged impingements and the inability of the conscious mind to deal effectively with death-related issues, the first need was to select for minds that could efficiently handle the excessive amount of incoming environmental stimuli and meanings. The solution opted for by natural selection had two critical and fateful components:

    a. *Selecting for subliminal or unconscious perception.* The first option in redesigning the emotion-processing mind was to select for minds that developed the capability of bypassing conscious perception and registration for a large part of the most disturbing emotionally charged events and meanings with which they were confronted.

i. This process was developed by means of an automatic, mental assessing operation, which gauged the extent of potential anxiety and disruption for the conscious mind that the meanings and implications of a given emotionally charged trigger event would entail.

ii. This assessment function, assigned to an *unconscious message-analysing centre of the emotion-processing mind* (Langs, 1986, 1987a, 1987b, 1988), can be thought of as a gating mechanism barring entry of potentially disruptive perceptions into the conscious system or as a selection mechanism for the overall emotion-processing mind.

iii. Unconscious perception spares the conscious system much of its emotional input load, so the design makes sense adaptively.

iv. By opting for input screening and selection mechanisms that direct incoming information and meaning towards one or the other system (conscious or unconscious), the mind could be arranged so that the conscious system preferentially received information and meaning that were most vital to survival—physical dangers, significant psychological dangers, and the like. The remainder—unimportant noise and strong emotional inputs of the disruptive variety—could be directed to a second system of the mind via unconscious or subliminal perception and processed therein. A mixture of universal general rules and specific individual sensitivities prevailed.

v. There are evident benefits to a two-system mind by means of which adaptation can take place on two levels simultaneously.

vi. The division of labour also seems advantageous—one system sharply focused on real traumas and dangers, the other concentrating on emotional concerns (including the emotional impact of frame-related experiences).

vii. As for the overall cost and disadvantages of this evidently favourable design decision, the selection for conscious defence over conscious registration with subsequent muting proves to be expensive emotionally. The absence of conscious registration of important emotional meanings inherent to trigger events creates situations in which critical aspects of emotionally charged experiences fail to reach awareness, placing the individual in a distinctly disadvantageous position in terms of adapting to emotionally charged relationships and interactions.

In essence, the selected design greatly reduced the input load on awareness, and conscious-system functioning was preserved, but what was lost is critical to emotional life and its adaptations. This was indeed a major evolutionary trade-off, with distinct advantages and disadvantages.

b. *Selecting for conscious-system defences.* Complementing and supporting the selection for unconscious perception was a second option chosen by natural selection in its efforts to safeguard conscious-system functioning from system overload and from meanings that would lead it towards dysfunction. This choice entailed the development of very powerful forms of conscious-system defences, especially denial, repression, and displacement—another decision for defence rather than some form of protected awareness.

i. The conscious mind was equipped with strong capabilities for *denial* that were the other side of *unconscious registration*. The meanings and trigger events that are registered subliminally are automatically subjected to *conscious-system* obliteration. But, in addition, provision was made for a back-up defence in the form of the frequent use of *repression*, so that disturbing meanings that did register in awareness and were stored in the superficial unconscious memory subsystem of the conscious system could subsequently be kept from returning to awareness.

ii. Two forms of back-up repression and denial also

were favoured: first, a natural opposition to the very expression of encoded images or derivatives of unconscious meaning; second, a disinclination to engage in trigger-decoding and the linking process through which themes are connected to their adaptation-evoking trigger events to convert deep unconscious experience into *decoded*, conscious insights.

iii. We see, then, that several types of obliterating defences operate within the emotion-processing mind. The first are active on the perceptual level and involve a form of gating and non-perception through which critical meanings fail to gain direct access to awareness. This defence also ensures that conscious and deep unconscious experience will be very different and that the two systems will operate quite separately—adaptively and otherwise. The second group of obliterating defences follows the *conscious registration* of a painful input meaning. It entails a repressive barrier that stands as a defensive gradient between the memory bank within the superficial unconscious subsystem of the conscious mind and access to awareness.

iv. Work within empowered psychotherapy has shown unmistakeably that the conscious system stubbornly insists defensively on being unaware of many critical meanings inherent to emotionally charged trigger events. The system is also capable of repressing and/or denying perceptions and experiences—trigger events—that are more than obvious to a neutral observer and to psychotherapists trained in this mode of therapy. This is especially the case with frame-related triggers in that the meanings and effects of extremely threatening and harmful frame deviations typically fail to be noticed consciously. They are, however, registered unconsciously, even as they are, as a rule, kept in a lasting state of denial and repression by the conscious mind. As a result, a therapist like Dr March, alluded to in chapter 8, can damage his patient, Ms Wells, with a variety of frame deviations, yet none of this harm is experienced by her

144   EVOLUTIONARY SCENARIOS

(or by him) within awareness even though they influence her (and his) emotional state and life detrimentally.

2. *The processing-system problem.* While unconscious registration might well have been followed by the obliteration and loss of the meanings involved, natural selection did not opt for unconscious perceptions without processing. Instead, it selected for minds that developed the capability of *processing unconsciously perceived meanings outside of awareness.* In this way, a true two-system mind was created, and adaptive processing could be brought to bear on emotionally charged impingements on two levels: with and without awareness.

Several advances contributed to this solution, especially the development of a capacity for language-based internal representations that lend themselves to effective, intelligent, mental manipulation. As a representational system that processes direct and symbolic portrayals, the deep unconscious system could draw on earlier forms of representation and adaptation, such as the mimetic capabilities of *Homo erectus*, to which the more efficient language-based system was added.

While non-verbal forms of intelligence and adaptation continued to operate, they are far less powerful than the language–thought–symbolic–conceptual cognitive system that emerged with language acquisition. Natural selection therefore opted for an effective, language-based, second processing system of the emotion-processing mind. The result was a parallel-processing, two-system design that could handle, without going into a dysfunctional state of system overload, the huge amount of emotional stimuli that *Homo sapiens sapiens* needed to process each day.

   a. In the emotional domain, deep unconscious intelligence—the deep unconscious wisdom subsystem—proves to be a far more effective adaptive organization than conscious intelligence. Factors in this outcome seem to include the following:
      i. impairments in conscious adaptations caused by conscious awareness of competing options;

ii. the disturbances in emotionally relevant learning that arise from distractions and conflicts affecting the conscious system;
  iii. the defensive posture of the conscious mind, which interferes with gaining knowledge of the environment and of available adaptive options—*conscious-system* repressions and denials are costly in that they preclude or interfere with perception, knowledge acquisition, and learning;
  iv. the relative absence of defensive obliteration within the deep unconscious system, so that perception is incisive and learning can proceed relatively unencumbered;
  v. the self-punitive needs of the conscious system that are translated both into the selection of poor and self-destructive adaptive options and into learning errors.
b. We can see then that the defensiveness afforded to the conscious system of the emotion-processing mind protected its fragile adaptive efforts, but that it did so at great cost. The price includes the following:
  i. ignorance of critical inputs from self and others, and the environment as a whole;
  ii. conscious learning difficulties and therefore adaptive impairments and a strong inclination towards maladaptive responses to environmental impingements;
  iii. a lack of preparedness for hurtful impingements from others that could have been anticipated were it not for the denial and repressive defences of the conscious system;
  iv. considerable emotional suffering caused by *conscious-system* defences that render it impossible (without trigger-decoding) to be aware of the unconscious sources of much of our emotional conflict and pain;
  v. strong inclinations towards maladaptive responses to emotionally charged trigger events because the more

adaptive solutions are arrived at unconsciously and not passed on directly to awareness;

c. In opting for protecting the conscious system via unconscious perception and an array of psychic defences, natural selection has supported a design for the emotion-processing mind that renders *Homo sapiens sapiens* relatively vulnerable and inept—and self-defeating—in the emotional realm. This seemingly poor outcome of evolutionary testing and selecting may reflect:

i. The unavailability of more effective options, a possibility that speaks for the intensity of the seemingly unsolvable dilemmas that natural selection has had to deal with in fashioning the emotion-processing mind. This possibility also implies that the technological, cognitive, and cultural advances forged by *Homo sapiens sapiens* have far outrun its capabilities to deal with their many ramifications and consequences for emotional life in general. Maladaptations may become even more prominent and striking as future decades unfold. Small wonder that many anthropologists believe that, as a species, we are heading for self-made extinction (Leakey & Lewin, 1992; Ward, 1994).

ii. A more optimistic but still troubling possibility is that 150,000 years is simply too little evolutionary time for natural selection to have guided *Homo sapiens sapiens* towards a more effective and less costly design of the emotion-processing mind. We certainly know that it can takes millions of years to effect truly critical design changes in physical features, including the brain—much the same may well hold for psychological features as well. This makes it all the more urgent for us to use our intelligence, imagination, and creativity to find ways to compensate for and correct the evident design flaws and limitations of the emotion-processing mind, lest we as a species meet the fate of almost all the other species that have populated this earth. Indeed, the indications that we are moving in the wrong direction on these issues speak for the inadequacies of the architecture of the emotion-processing mind.

3. *The output problem.* We turn now to the question of how variation and natural selection designed the mental and behavioural channels of output from the deep unconscious system and its adaptive intelligence—the means of expression selected for this system. The main choices appear to be:
   a. allowing these processing efforts to go to waste without registration and expression of any kind;
   b. generating output without awareness in some language form and/or via somatic channels of expression;
   c. the forging of some type of communicative bridge from the deep unconscious wisdom subsystem to the conscious system of the mind.

   The main dilemma here can be simply stated: The deep unconscious system was designed to process, without awareness interceding, our most painful and emotionally disruptive inputs and their most terrifying meanings. Thus, opting for a direct output report on its operations would mean that the conscious mind would become aware of the very contents that the input and processing parts of the design of the emotion-processing mind had disallowed. On the other hand, opting to keep these processes and adaptive decisions outside of awareness would deprive the individual of profoundly important perceptions of the environment and the sound adaptive solutions reached by the deep unconscious wisdom subsystem. Given the intensity of this conflict, what then did variation and natural selection combine to produce?

   a. *Selecting for the encoded communication of the results of deep unconscious wisdom subsystem processing.* The choice made by natural selection with regard to outputs from the deep unconscious wisdom subsystem was once again highly defensive and partially, though not entirely, obliterating. We cannot help but be impressed with natural selection's insistence on defensive and obliterating choices for the design of the emotion-processing mind at virtually every turn. The output features of this design include the following:

i. The absence of direct reports to the conscious mind of the unconscious perceptions and processing results of the deep unconscious wisdom subsystem's responses to emotionally charged trigger events.

ii. However, allowance was made for *encoded reports* on these processes. As a result, *Homo sapiens sapiens is a natural-born encoder* and instinctively or psychobiologically inclined to use narrative communication to encode its deep unconscious perceptions and adaptive recommendations.

iii. Without direct awareness of the results of the unconscious processing of subliminally perceived experiences and their meanings, the conscious system—and the total human organism—is deprived of its most intelligent and adaptive responses to emotionally charged impingements.

Summing up

It is evident that natural selection was willing to pay dearly for protecting the survival functions of the conscious system of the emotion-processing mind. The enormity of the cost for these input and output defences suggests that conscious functioning is not only vital to survival, but an extremely delicate adaptive system that is all too easily disrupted. The choices made also imply that allowing distressing, unconsciously experienced meanings access to awareness is a far greater risk to survival than obliterating these meanings and with them the important information and insights that they contain.

The results are something less than "satisficing" (Simon, 1962). Indeed, the basic predicament does not seem to have been resolved. It is still the case that if we obliterate the meanings of emotionally charged experiences, we suffer, but if we know what is going on, we evidently suffer even more. Natural selection has an enormous amount of work yet to do to improve on the design of the emotion-processing mind. The final word for the moment appears to be that, in this respect, we are—quite unwittingly—part of one of nature's most awesome and awful experiments. It would seem that if we do not take over this design problem from nature and change the workings of this cognitive mental module ourselves, we are likely to become the victims of one of natural selections most compromised, if not failed, engineering jobs—designing the emotion-processing mind.

CHAPTER TWELVE

Solving the problems of death and violence

We have been exploring the evolution of the processing or adaptive systems of the emotion-processing mind of *Homo sapiens sapiens* primarily in terms of the issue of conscious system overload caused by the massive selection pressures that arose from the enormity of its social and other environmental impingements. There was, however, a particular set of issues among these impingements that appears to have been of overwhelming importance and responsible for a significant segment of the processing pressures placed on the conscious system of the mind of this species.

1. The first of these issues is a problem that was faced by a number of earlier species—that of violence from conspecifics. Danger from predators drawn from other species exists on virtually every level of species development; in *Homo sapiens sapiens*, dealing with this problem primarily is the province of the conscious system. But conspecific violence, which includes the killing of infants and young children by parents and extends to adult violence of all kinds, has escalated among later animal species. Even so, *Homo sapiens sapiens*

appears to have considerably expanded the range and degree of conspecific violence to an extremely dangerous extent (Alexander, 1989). Finding the means of managing and controlling these impulses and actions therefore soon became another critical selection pressure for natural selection in choosing features for the emotion-processing mind.

   a. A subsidiary and related issue pertains to conspecific incest and other inappropriate sexual liaisons. While protection against incest and other sexual problems seems to have been genetically programmed into almost all mammalian species (Brown, 1991), natural selection needed to ensure the enforcement of these constraints and barriers for *Homo sapiens sapiens* as well. In general, humans do not and cannot depend entirely on genetically fixed defences for their culturally designed prohibitions, so additional safeguards needed to be selected for and incorporated into the design of the emotion-processing mind.

2. The second major issue that overtaxed the conscious mind is that of a definitive awareness of personal mortality. As noted earlier, language acquisition brought with it a strong sense of personal identity and a sharp distinction between self and other, as well as an elaborate capacity to anticipate the future—including the inevitability of the death of self and others, and a full awareness of incidents in which death is pending or has occurred.

This kind of awareness had mixed effects on adaptive coping by *Homo sapiens sapiens*. On the one hand, it sharpened realizations of the possibility of annihilation. This led to advances in personal and collective protective mechanisms and improvement in the capacity to anticipate dangerous relationships and situations, thereby promoting personal safety. But on the other side there was, firstly, the development of unbearable existential anxieties that needed to be reduced or controlled, and, secondly, an expansion of weaponry and an intensification of attacks on others—carried out in part to deny a given individual's own consciously and unconsciously realized vulnerabilities. This created a vicious cycle in which aggression and violence kept escalating—a problem that needed a more effective emotion-processing mind for its resolution, lest conspecific annihilation materialize for our entire species.

The increase in attacks on others evidently arose both as a way of protecting oneself from an anticipated attack by someone else and as a deeply motivated means of denying personal helplessness in the face of the inevitability of one's own death—an attempt to create a denial-based, delusional belief in one's own immortality. Repeatedly, one complex selection pressure interacts with other complex selection pressures to create grave challenges for the engineering capabilities of natural selection. In light of the complexity of the life and environments of *Homo sapiens sapiens*, the *evolution of selection pressures* has presented a number of unsolved challenges for evolutionary change that have yet to be met.

The evolved processing solution

Natural selection needed to test out and sort through a wide range of variations in order to select for the most adaptive means of solving these two major problems—conspecific violence and existential death anxiety. As a problem in engineering, there appear to be two possible means of recourse:

1. Adapting to these selection pressures through the choices made for the basic information- and meaning-processing ca-pabilities of the emotion-processing mind. This is *the basic design solution* in that death anxiety and issues of violence bring many psychodynamic forces into play, and adapting to these concerns undoubtedly has had effects on the selected universal design of the emotion-processing mind. For example, these adaptive problems are replete with disturbing perceptions and meanings; needing to cope with these inputs undoubtedly contributed to the development of a two-system mind and to the pervasive use of denial and repression by the conscious system.
2. The second option is to create a new system of the mind to facilitate adaptation to these two ominous adaptive issues.

DEATH AND VIOLENCE   153

In dealing with violence and death anxiety, natural selection opted for both of these possibilities.

1. *The basic design solution*

As for the first of these choices, let us look more carefully at how these two dynamic problems helped to shape and are dealt with by the conscious and unconscious processing systems of the emotion-processing mind.

   a. The processing problem once more presented natural selection with conflicting choices, a striking characteristic of testing and selection for the emotion-processing mind. With regard to violence, the available options are, on the one hand, heightening the awareness of threat from others, which could well escalate violent responsiveness and create a vicious, circular interaction of increasing warfare between conspecifics, or, on the other hand, opting to obliterate many of the psychological meanings of aggressive environmental impingements—a choice that would increase an individual's vulnerability to conspecific attack. Once more natural selection stood between a rock and a hard place—there is no clear line to the most effective and least costly solution to these dilemmas. The options seem to entail either rendering us overly sensitive to threat and thereby excessively assaultive or making us overly insensitive to threat and inclined to be victimized and harmed.

      i. Obliterating types of conscious defences would be disadvantageous because subtle hostile communications from, and threatening behaviours by, others would often be missed consciously. In addition to rendering us vulnerable to unexpected attacks that could have been foreseen, this arrangement would cause a build-up of repressed and denied images of threat and danger that could culminate in unconsciously motivated protective and retaliatory actions that are released in an explosive manner because the provocative in-coming perceptions have been intensifying yet are unavailable to conscious processing.

ii. The processing solution chosen by natural selection regarding hostile inputs from the social environment once more, true to form, leaned towards obliteration rather than heightened sensitivity. As we saw in the previous chapter, the emotion-processing mind has been selected largely with conscious defensiveness as the prime choice.

iii. In substance, then, by virtue of natural selection, an enormous amount of threat, hostility, and pending danger from others does not reach awareness, and in general we are rendered naturally vulnerable to environmental threats. For example, a considerable amount of damage is done to us by those who impose frame violations on us, yet we fail to register consciously the true sources of the hurt and harm caused by these violations, even as they detrimentally affect our lives. The blind acceptance of considerable personal harm is a major price we pay for many of the protective aspects of the design of the emotion-processing mind.

iv. As for death anxiety and the reality of human mortality, the design of the emotion-processing mind has similarly opted for defensiveness rather than heightened awareness. Both denial and repression play a notable role in this alignment, as does the assignment of many death-related experiences and meanings to processing by the deep unconscious system. Indeed, a remarkable proportion of death-related experiences and meaning does not register in awareness.

v. Basically, the human mind is designed so that responses to death and threats of loss through death receive a highly constricted and limited response from the conscious system. A mild illness in oneself or a loved one will be dismissed consciously, while it is simultaneously worked over extensively in the deep unconscious system. This is the case for all sorts of reminders of a given individual's mortality. In this way, we are protected from many moments of overwhelming anxiety. But we remain vulnerable to the unconscious effects of these death-related realizations, which strongly and unknowingly influence our life choices, daily behaviours, and feeling states.

vi. The main adaptive defence against realizations related to the inevitability of the death of loved ones and self is that of *denial* (Becker, 1973). Nature does not appear as yet to have produced significant variants on adaptations in response to death anxiety, and denial seems to be all we have. Still, it is well to remember that variation and natural selection have had only 150,000 years to work on developing an optimal cognitive, language-based survival strategy in response to this unprecedented selection pressure.

vii. To appreciate the pervasiveness of denial mechanisms invoked to deal with death-related concerns and threats, we need to recognize the many forms that denial of death takes in addition to obliterating psychologically one's awareness of the realities of death itself. Among these denial equivalents there are manic flights of ecstasy and celebration; unusual feats of power, often death-defying in their nature; and the use of drugs and alcohol.

Less well known as a denial-of-death mechanism is the invocation of frame breaks—violations of rules, boundaries, and frames—which serve to deny the existential rule of life–which is, of course, that it must end in death. Frame deviations are perhaps the most common means through which humans attempt to deny the inevitability of personal demise. As such, the inclination to modify frames is virtually built into the design of the conscious system, which tends to adapt to death-related issues through the invocation of denial defences rather than confront and deal with them head-on.

viii. The denial of death has its benefit side in relieving us from undue anxiety and a preoccupation with the inevitability of death that could derail *conscious-system* functioning. But what, then, is the price we pay for these defensive denial mechanisms?

Denial of death often interferes with seeking proper medical care when there are signs of a dangerous illness. In addition, the use of psychological denial mechanisms can be distinctly maladaptive in creating flights from necessary responsibilities, hurtful forms of promiscuity, and inappropriate exploitation of, and assaults on, others, including outright acts of uncalled-for violence. Denial through frame alterations harms both the frame breaker and the people who are victimized by the frame break, doing so through deceptions, disloyalties, abuse, inappropriate seductions, abandonments, hurt, and the like.

Denial is an expensive means of defending against personal death and loss. When denial is utilized, learning and adaptive change are all but impossible, and the actions taken to support the denial defence typically are harmful to self and others. We await the moment when an unprecedented and ingenious way of dealing with death anxiety emerges as a variant and can only hope that the situation in which it materializes will allow for its favoured selection and reproduction. Until then, once more we have much intellectually based work to do in order to counter and overcome the flaws in the architecture of the emotion-processing mind with respect to its adaptations to death-related trigger events and their ramifications.

2. The creation of a new system of the mind

The evolved architecture of the emotion-processing mind appears to offer very little in the way of sound resolutions to the problem of conspecific aggression and violence. Yet these issues in their varied expressions undoubtedly are one of the greatest existing threats to the survival of the *Homo sapiens sapiens* species. Natural selection seems to have recognized the crisis nature of the situation and opted to select for minds that fashioned a second deep unconscious subsystem—*the fear–guilt subsystem*. This system embodies some fresh and creative ways of dealing with both the fear of death and conspecific violence. Let us look at the structure and functions of this relatively new subsystem of the mind.

   a. The fear–guilt subsystem activates fears of personal death and guilt for transgressions of rules, frames, and boundaries, and for all types of harmful acts and impulses—fantasies, conscious and unconscious, directed towards others and oneself.

b. As for death anxiety, survival is served by storing the fear of death in an unconscious system of the mind and restricting its conscious realization to times of acute danger and moments of strong death-related tensions and threat. Adaptation is also served by keeping these potentially disruptive concerns from conscious awareness. In addition, the fear of death in the form of personal harm by others is part of our evolved means of controlling our own destructive impulses towards them. While consciously we are well aware that attacking someone risks his or her reprisals, unconsciously the fear of death serves also to constrain our impulses to harm them.

c. But, as we have come to expect, there is a price to be paid for this protection and for the adaptive value of locating much of our fear of personal annihilation in the deep unconscious system. The fear of death is ever-present, and, because it operates largely outside of awareness, it influences our behaviours, emotions, feeling states, inner mental life, relationships, and mental health, without our realizing its existence as a major motivating and causative force.

d. The dread of personal death is a strong motivator for the denial defences that are directed against death anxiety (see above). Unconsciously driven by the dread of our own mortality, we turn to violent and seductive frame breaks to delude ourselves into the belief that we are more powerful than death itself, and that we are the exception to the rule that death follows life. We seek out, believe, and behave according to ideas of life after death, often with striking detrimental consequences for ourselves and others.

e. Quite unexpectedly, we find that unconscious knowledge does not affect conscious adaptations, but the unconscious fear of death greatly influences our conscious actions and choices. Unconscious knowledge is experienced as dangerous, while the unconscious fear of death is protective—even though it is also a dangerous motivating force.

f. What, then, of the guilt component of the fear–guilt subsystem? In that regard, we find that humans are genetically programmed and learn cognitively that hurtful transgressions do and should evoke feelings of remorse and guilt, and a measure of self-attack. There is, of course, *conscious guilt*, which is activated when acts of harm are recognized in awareness and which constrains some of our impulses to harm others physically or psychologically. But conscious guilt tends to be weak at times of threat and in the face of impulses for revenge, urges that have intensified in *Homo sapiens sapiens* because of the escalation of social pressures and the personal awareness of helplessness in the face of death and other overwhelming situations.

Additionally, many profoundly harmful actions taken against humans by other humans unfold without either the conscious realization that harm is being done or the activation of conscious guilt. In these instances, *unconscious guilt* is needed as an added restraint; it is activated in these circumstances and as a preventative as well. Unconscious guilt strongly influences conscious-system behaviours, largely by leading humans to seek out unwittingly self-harm and self-punishment as ways of curtailing inclinations towards conspecific violence.

g. There are two evolved reasons why *unconscious guilt* is such a powerful influence on the conscious system:

   i. The first is that consciously experienced guilt and remorse is painful and diverting, and humans tend to avoid it or rationalize it away; it is a weak restraint and does not effectively contain human impulses to harm others.

   ii. The second reason for the evolution of unconscious guilt lies with its much-needed power to constrain an individual's unconsciously driven tendencies towards violence and the harm of others. By selecting for minds that are continuously unconsciously motivated to experience and satisfy needs for punishment when inappropriate violence and sex and other frame violations occur, natural selection seems to have found a means of reducing violence among fellow members of the *Homo sapiens sapiens* species. The benefit from this arrangement is a lessening of assaults on others, but the cost is found in our frequent acts of self-punishment and self-harm. A large number of maladaptively self-defeating behaviours and choices are made on the basis of this aspect of the design of the emotion-processing mind.

All in all, the most cursory survey of the conditions of the world today suggests that the selection of design features for the emotion-processing mind with respect to adapting to personal death and for controlling conspecific violence have been far from successful. We are as a result in possession of badly compromised minds when it comes to processing and adapting to fears of death and inclinations towards violence against others. Nowhere is there a more pressing need than here for finding intelligent ways to improve on the work of nature's hands—our personal survival and the survival of our species may well depend on it.

CHAPTER THIRTEEN

The evolution of frames, rules, and boundaries

We have one more evolutionary puzzle to sort out in this pursuit of a comprehensive adaptationist programme for the emotion-processing mind. We have seen that the two systems of the mind are designed to respond very differently to impingements that involve rules, frames, and boundaries. These differences, and the basic importance of human attitudes and behaviours as they pertain to boundary and frame factors, render this aspect of the evolved design of the emotion-processing mind both intriguing and of considerable import. It therefore behoves us to take a closer look at this aspect of human functioning and to try to develop an adaptationist programme for the evolution of the frame-related adaptations, choices, and behaviours of *Homo sapiens sapiens*.

Why frames are important

There are several reasons why frames, rules, and boundaries are critical factors in emotional adaptation and why this aspect of human life is the prime concern of the deep unconscious system of the emotion-processing mind. To list them here, they are:

1. In all of nature, both physical and biological, *boundary conditions* define and demark entities. For biological entities, intact boundaries are essential for their survival, and transactions at the boundaries are critical to the functioning and adaptations of the organism and its subparts. Many of the metabolic activities of biological entities take place at the boundary, including the intake of energy and other vitally needed substances and the excretion of wastes (de Duve, 1995). In addition, boundaries are the contact points between entities and their environments, including other entities. Damaged boundaries usually lead to dysfunctions in, or the destruction of, an organism or one of its subparts. The human skin and the cell membrane are prime examples of boundaries, and they indicate how boundaries are essential for the definition and survival of organisms.

    Activities at the boundaries of biological entities are also a vital source of information and meaning, and they greatly affect the functioning of the total system. The determinants of what does and does not pass through the boundaries of an entity are crucial to its functioning. In addition, in humans, boundary conditions are a critical factor in shaping and determining the meanings of expressed communications and the outcome of interactions and processes that take place within a given space and/or system, as seen in a patient/therapist system within a therapy space.

    Another notable example of boundaries relevant to psychotherapy includes the postulated boundary that protects the conscious system by screening out disturbing information and meaning, a boundary function that separates the conscious and deep unconscious systems of the emotion-processing mind, rendering them relatively independent entities. This particular boundary is an essential feature of the emotion-processing mind and is a factor in defining its structure, functions, and adaptational capacities. On a very different level, the physical structure of a therapist's office is a different type of boundary condition with many evident and more subtle ramifications.

2. Similarly, *rules and regulations* are also essential for the definition and integrity of a system and for its ordered and effective functioning. While rule modifications do provide opportunities for creative adaptations and fresh solutions to dilemmas faced by the emotion-processing mind, stable rules properly govern most daily functions and ensure their smooth and effective continuation and operation and also serve as a much-needed backdrop for innovative, constructive rule changes.

Rules operate on many biological levels, ranging from DNA and cells to the total individual organism, and, in humans, they apply to both the physical and psychological realms. Rules are essential for the smooth operations of the emotion-processing mind, as well as for society and culture. In all, then, both rules and boundaries are critical dimensions of human life.

3. The design of the emotion-processing mind as it applies to dealing with rules, frames, and boundaries is unexpected and in need of explanation.
    a. Despite the importance of frame impingements, the *conscious system* is inclined to ignore them. This obliterating position is a design feature that is fateful for human life and for psychotherapy, in which both patients and therapists tend to ignore ground-rule and setting activities despite their pervasive effects on both parties to treatment. As a result, the negative consequences of *frame deviations* seldom register in conscious awareness.
    b. In addition, despite the value and importance of *intact or secured frames and boundaries*, paradoxically, the conscious system favours ruptured, damaged, and deviant or unsecured frames.
4. The deep unconscious system is exceedingly responsive to frame impingements and is a frame-centred system of the mind. It also consistently advocates and favours frame-securing behaviours and interventions.
    a. While this is in keeping with biological expectations, the enigma lies with the finding that these highly salutary attitudes and preferences are inaccessible to awareness

and do not affect human adaptive responses to frame-related events.

We are faced with a number of seemingly inexplicable anomalies and unsolved puzzles. Let us see what an evolutionary perspective can offer by way of clarification.

The basic functions of frames

As noted, frames in the form of rules and boundaries play a crucial role in the functioning of individual organisms, their subsystems, and their interactions with their environments. As such, they are the guarantors of order and lawfulness, and the foundation on which disorder and inventiveness, and evolution itself, are based. Without rules and boundaries there is anarchy, utter disorder, and eventual annihilation.

Throughout biological nature, rules, frames, and boundaries serve the following functions:

1. The demarcation of individual entities and their settings.
2. A means of regulating, governing, and constraining interactions on all levels of organization—within organisms, between organisms, and within families and societies of organisms.
3. A means of restricting inappropriate and disadvantageous sexual liaisons.
4. A major constraint on inter-species and intra-species violence.
5. A fundamental means of providing an organism with a relatively stable *background* setting and set of guidelines for the transaction of highly changeable and often uncertain *foreground* interactions and events.
6. The protection of energy supplies and shelter, and other requirements for surviving on a daily basis.
7. A guarantor of reproductive opportunities and success.
8. A determinant of the nature of interactions and relationships,

and of the meanings and implications of behaviours and communications.

Frames, rules, and boundaries are *relatively stable entities*; in general, they change slowly with time. They form a backdrop or framework for negotiating the *ever-changing vicissitudes* of everyday life related to people, places, events, adaptive challenges, psychodynamics, interpersonal issues, and just about everything *Homo sapiens sapiens* must deal with and adapt to from moment to moment.

Given the changeability and uncertainty of present-day hominid life, frames serve as an anchor in a sea of emotional storms. While instability can, of course, have effects that range from chaos and disruption of adaptive functioning to highly innovative adaptive discoveries, humans need a stable set of rules, settings, and boundaries for inherent support in the face of foreground uncertainties. This relative stability of frames and boundaries is evidently very much appreciated and valued by the deep unconscious system of the emotion-processing mind.

In the research we carried out on communication in psychotherapy sessions and in every-day dialogues (Langs, 1992c; Langs & Badalamenti, 1992a, 1992b, 1994a, 1994b, in press), we discovered deep stabilities and regularities with regard to speaker duration (who spoke in sequence and for how long) and the shift in and out of narrative forms of communication. We also unearthed a set of consistent, deep laws related to complexity, work, and force that governs the use of *communicative vehicles*.

This deep stability seems to provide a foundation for the variability, if not the utter disorder, reflected in the surface exchanges between patients and therapists, and couples in dialogues. Natural selection appears to favour a strongly secured and stable platform of deep laws, regularities, boundaries, and rules as a basis for more variable surface transactions. Order, then, is needed to allow for both security and disorder, much as consistency is needed to allow for innovation and change.

Rules and boundaries have existed since virtually the beginnings of life (de Duve, 1995). Let us look now at their evolutionary history as it pertains to the emotion-processing mind and human emotional life.

An evolutionary history

All organisms survive through in-built and learned regularities and by virtue of some degree of physical and social organization, however crude. Many birds and vertebrates are territorial, with clear and generally respected boundaries that demark their living and breeding spaces. Within these confines, they are spared attack and threat from conspecifics and, in some cases, from predators from other species. They thereby safeguard their food sources and energy supplies and enhance their chances for survival and reproductive success (Griffin, 1984).

An incest taboo appears to be wired genetically and expressed instinctively in most vertebrate species, and this may well be the case for *Homo sapiens sapiens* as well (Brown, 1991). In addition, dominance hierarchies are common among vertebrates, including apes, and serve as a way of effecting social order. *Natural selection has favoured the use of rules and boundaries as protective and survival mechanisms of the first order.*

With the arrival of the hominids, interpersonal boundaries and rules pertaining to hordes, clans, and families expanded considerably. Social structure and elaborate cultures were developed via rule-bound constraints and well defined laws of interaction and behaviour. Laws designed to constrain incest and violence were put into effect. Social hierarchies were developed, with rulers and subjects, and a variety of additional well-defined and well-regulated social configurations. Rules, frames, and boundaries governed family and social relationships and interactions, including those pertaining to such vital activities as pedagogy, schooling, and work.

With the evolution of *Homo sapiens sapiens*, the constructive, enhancing, and safeguarding functions of rules, frames, and boundaries reached new heights of intensity, complexity, and functional utility. However, there is a critical emergent aspect of frames that arose through language formation and the awareness of personal mortality that is not in evidence in other animals. Once more, language acquisition introduced a new dimension into the adaptations of biological species, and *frame-related experiences took on unprecedented meanings.*

The safety-enhancing functions of rules, frames, and boundaries remain in place for *Homo sapiens sapiens*. However, this species alone is aware of personal mortality, and this awareness is unconsciously constructed in terms of frames and boundaries in two ways:

1. The basic existential rule of human existence is that life is framed and bounded by non-existence on one side and death on the other.
2. The frame of life is experienced as a closed space, a claustrum, which is entered at birth and from which there is no exit except through personal demise. Closed spaces and therefore secured, constraining frames have come to represent symbolically the entrapment aspect of the frame of life, and they are associated with the unconscious experience of the hopelessness of any escape from enclosures except through annihilation.

The result is that *secured frames*, which are experienced as essentially safe and protective in non-humans, are experienced in two contradictory ways by humans alone: first, as safe and enhancing, and, second, as entrapping and dangerous and linked with personal death. Essentially, secured frames are experienced as life-giving and death-giving all at once. Secured frames speak for the maxim that *the gift of life comes at the price of death*.

Given that death anxiety is the deepest and most powerful form of dread experienced by humans, the *unconscious experience* of secured frames involves both the danger of death and the enhancement of life. *Secured-frame anxieties* are the unappreciated source of the greatest dread that human beings experience, and these anxieties exert powerful, unconscious effects on everyone's lives. These anxieties are much like a silent plague that engineers all manner of diseased and maladaptive actions and behaviours whose deepest sources remain unrecognized.

For example, virtually everything that is maladaptive and hurtful about today's practices of psychotherapy has as one root the secured frame anxieties of both patients and therapists. It is this hidden terror that so strongly motivates ground-rule deviations in therapy situations and in life in general—the hunger for death-denying frame breaks is insatiable. There is a universal unconscious delusion in humans that a frame breaker can defy the existential rule that life is followed by death. Untold amounts of harm and damage have resulted from the dread of secured frames, even as they offer the most powerfully constructive settings and rules for our daily functioning and adapting—and for psychotherapy and our lives in general.

Natural selection seems stuck on the problem of how to deal with the awareness in *Homo sapiens sapiens* of personal mortality. As noted, it has not been able to find an effective variant other than denial in its myriad forms—including, of course, frame violations—to deal with this stark inevitability. But it can truly be said that the price we have paid and are paying for these denial defences and frame breaks is monumental.

With few exceptions, every murder, war, social crime, psychotherapeutic crisis, and human-caused disaster has at least one source in secured-frame anxieties, as does almost every personal hurt and intrapsychic and interpersonal conflict. *Homo sapiens sapiens* has not found an effective adaptation to the prospect of dying—and we are paying dearly for this failure. The incredibly adaptation-enhancing gift of language has brought with it an equally incredible cost in personal and collective dysfunction and harm that serves as a grim reminder that evolution and natural selection are quite imperfect—especially in the short term.

CHAPTER FOURTEEN

Assessing the accomplishments of evolution

I have offered a series of adaptationist programmes that pertain to the evolved structure and functioning of the emotion-processing mind. We repeatedly found that as we moved along, selection pressures became more and more intense, the choices of variants more and more constrained, and the available time period to effect evolutionary change less and less. As a result, our scenarios led us to the discovery of choices made through natural selection that seemed gravely compromised if not essentially dysfunctional.

All in all, it appears that the human mind is not only one of the most energically costly entities in the human organism, but also one of the most expensive and poorly honed adaptive organs in the human cognitive armamentarium—even as it is perhaps our single most critical and vital asset and means of coping in the emotional realm. The very emotions that give richness and meaning to our lives prove to be our least understood and most costly adaptive issue.

This state of affairs arose because natural selection was faced with many unfamiliar dilemmas and conflicting choices in selecting for the emotion-processing mind, and it was compelled to operate without clearly defined cost-benefit ratios. *Homo sapiens sapiens* evolved to a point where social pressures and language capacities greatly increased the power and impact of emotionally charged relationships and interactions—so much so that these impingements began to overwhelm the mental apparatus. Until then, the more knowledge an organism had of its environment, the greater its chances of survival. Suddenly, knowledge, one of the most valuable biological assets, became a liability and danger—a burden and threat to survival. Natural selection had a remarkably unprecedented set of selection pressures to deal with—and little time in which to do so.

A major factor in this problem is the fact that the key evolutionary development for hominids, language acquisition, is only some 150,000 or so years old—a remarkably short span of evolutionary time. This realization alone provides us with an invaluable perspective into many of the unexpected oddities and costly design features of the emotion-processing mind.

I turned to evolutionary psychoanalysis because there were design features of the emotion-processing mind that seemed to be inexplicable in light of the prevailing theories developed at other levels of the hierarchy of psychoanalytic science. In chapters 7 and 8, I defined some of the more mysterious and anomalous features of this mental module. Let us return briefly to these unexpected attributes in an attempt to gauge the success of the adaptationist programmes that I have proposed and to see if this final summing up will shed still more light on the relevant architectural issues.

1. The existence of two discontinuous processing mental systems seems, in its essence, to have been selected to keep our most terrifying and disturbing perceptions of others and ourselves from reaching awareness and disrupting *conscious-system* functioning. This arrangement follows two design principles:
    a. When a processing system reaches the point of information and meaning overload, a second system is created to handle the excess.
    b. When a single system must carry out contradictory functions—here, being both aware and unaware—a second

system is created to handle one of these functions, and the other function is left with the original system.

2. Selecting for *defensive protective mechanisms* built into the basic structure of the emotion-processing mind was explained as natural selection's main means of safeguarding *conscious-system* adaptive functioning and protecting it from system overload. This evolved design ensured smooth conscious-mind, survival-related operations but has proven to be very costly in loss of knowledge of the environment and in the automatic use of behavioural *displacements*. This latter mechanism helps to protect the conscious system from being aware of many disruptive contents and events, but the natural tendency to work over emotional issues both unconsciously and in arenas and relationships other than the actual source of conflict leads humans to engage in an enormous number of behaviours that are unwittingly dislocated from their true origins and therefore misapplied and maladaptive. We are remarkably unaware of the deep motives related to much of our emotional lives—we do much on the basis of reasons and motives that are far from our conscious rationalizations.

3. The power of the deep unconscious fear–guilt subsystem over the actions of the conscious system was explained primarily as a means of buttressing our weak conscious guilt system, and as a way of managing the problem of unconsciously motivated, uncontrolled aggression and conspecific violence, which appears to be a major unresolved problem in adaptation for *Homo sapiens sapiens*.

   a. The placement of our most intense fears of personal death in this deep unconscious system was explained as a means of reducing death anxiety, one of our most significant sources of dread and potential dysfunction.

4. The absence of tracts from the deep unconscious wisdom subsystem to the conscious mind found its evolutionary explanation in the now familiar hypothesis of the preference for obliterating defences rather than allowing for the conscious registration of many of the most painful meanings of our perceptions and experiences of others and ourselves.

5. The preference of the conscious system for frame deviations

found its clarification in the ways in which these frame modifications serve to deny death anxiety and neutralize, though at great cost, secured-frame anxieties.
6. The focus on frames and the preference for secured frames found in the deep unconscious system was seen to be a sensitive reaction to the critical role in adaptation and life played by inherently enhancing, yet terrifying, secured-frame conditions.
7. The choice of defence rather than enhanced conscious perceptiveness found part of its answer in natural selection's inability to find any means other than *denial* to deal with the prospect of personal death. This choice was also based on a related inability of the fragile conscious system to function well in the face of an awareness of the more disruptive meanings of emotionally charged trigger events related to the experience of the behaviours and communications from others and oneself.
8. A similar explanation was used to account for the intense opposition of the conscious system against direct access to processed unconscious meanings and even to their encoded expression.

In recounting the evident basis for these puzzling attributes of the emotion-processing mind, it is well to pause with one of the most compelling insights to stem from these evolutionary explorations. This refers to the existence of a powerful and deeply insightful deep unconscious wisdom subsystem, whose adaptive processing and directives for adaptive solutions to environmental problems find no direct access to awareness and have virtually no effect on an individual's adaptive responses.

This defensive design-feature entails a great sacrifice of a critical adaptive resource. But evolution via natural selection has a way of developing faculties stepwise. Thus, it may well be that even without access to awareness, the possession by *Homo sapiens sapiens* of a deep unconscious processing system is a first step towards an eventual design of the emotion-processing mind that will allow humans to benefit from the great adaptive wisdom embedded in the deep structure of their own minds. This kind of slow building process is not uncommon in nature, though it takes a great deal of time to bear fruit.

Possible new designs

Clearly, we are in need of a better design for the emotion-processing mind—the present one is too costly and may, as noted, lead to our extinction. On the whole, natural selection has favoured protecting the conscious mind through the massive use of perceptual and psychological defences—it has been inclined towards defence rather than truth.

The key question is this: is this arrangement a stop-gap measure along the way to a more effective design for the emotion-processing mind, or will it remain fixed for thousands or millions of years and contribute to the growing risk of extinction faced by *Homo sapiens sapiens*?

Although quite uncertain—engineering (imagining or creating a design) is far more difficult than reverse engineering (analysing an existing design; see Dennett, 1995)—it seems of value to offer a few speculative comments on the possible future design of the emotion-processing mind. I will do so mainly as a way of trying to discover ameliorative that we can engineer at the present time through our gift of human ingenuity—a knowledge heuristic that can, indeed, free us from the constraints of genetically founded, slowly changing design features.

The following possible revisions in the architecture of the emotion-processing mind would, I think, enhance our species' chances of survival and reproductive success—and improve the quality of our lives as well.

1. The initial improvements probably should involve the conscious system. This could include:
    a. Providing the conscious system with the capacity for multiple simultaneous registrations of incoming information and meaning, and with a capacity, similar to that of the deep unconscious system, to process multiple triggers and their various meanings simultaneously—that is, to free the conscious system from the limitations of serial processing.
    b. Finding the means of allowing the conscious system to tolerate presently horrifying perceptions of others and self, so that vastly more emotionally charged information

and meaning could, without dysfunction, be processed within awareness.
   i. This implies a strengthening of the conscious mind so that its adaptive functioning is far less easily disturbed than it is at present.
   ii. Also implied is a great increase in the emotional load capacity of this system.
   iii. Another implication involves the enhancement of *conscious-system* intelligence so that these advances are utilized to increase survival and reproductive success.
   c. Enabling the conscious mind to find new and especially effective ways of dealing with death anxiety.
      i. This implies a significant reduction in *conscious-system* secured-frame anxieties, a step that would lessen the extent to which we turn to harmful frame deviations.
   d. Finding a truly effective way of greatly reducing conspecific violence to improve our chances, as a species, for long-term survival.
2. The deep unconscious system would be reduced in size and become less vital for human survival. Much of what is processed in that system would be processed in the conscious system. Alternatively, the processing efforts of this system would have direct access, without undue disruption, to the conscious mind and its direct adaptations.

The means of change

Accomplishing any or all of this through variation and natural selection is a difficult and exceedingly slow proposition. Thus, the situation calls for actions that we ourselves can take at the present time using our intellect and secondary and tertiary heuristics (Plotkin, 1994) to enhance the existing adaptive capabilities of the emotion-processing mind. The programme through which this may be carried out includes at least the following measures:

1. Recognizing that the conscious mind is fundamentally antagonistic towards and defended against deep unconscious perceptions and realizations.
    a. This is a design problem and, as such, is difficult to overcome. Only intense conscious awareness of the inability of the conscious mind to allow deep unconscious experience take hold in awareness can keep us on track here. Once this crucial problem is recognized, measures can be taken to lessen the intensity of, or to bypass, these defences.
2. Overcoming resistances to trigger-decoding, the only means through which we can access our own deep wisdom and best adaptive solutions. Here, too, conscious-system obstacles must be recognized and modified.
3. Developing a strong and unswerving motivation to access deep unconscious meaning. We must come to understand truly that deep unconscious perceptions, forces, and motives are the strongest factors in our emotional lives and that, without awareness of unconscious experience, we are bound to suffer because the fear–guilt subsystem system rather than the deep unconscious wisdom subsystem has the natural power to affect the conscious system and its efforts at adaptations.
4. Appreciating the damage done by frame deviations on the one hand, and the healing powers of secured frames on the other. Once we can both tolerate death anxiety and control our tendencies towards conspecific violence better, and are better able to handle consciously the awareness of exceedingly painful perceptions of others and ourselves, our frame preferences are certain to improve.
5. Seeing to it that these insights become part of the fabric of today's practice of psychotherapy and the psychoanalytic theory on which it founded.

Concluding comments

I conclude this solemn discussion with some words of hope. We are designed by natural selection to deny death, and our supplementary intelligence, personal and cultural, has not improved on this denial. Instead, individuals and cultures concentrate their efforts on enhancing denial mechanisms by measures that include religious beliefs, ideas of life after death, and even mass suicides. These adaptive efforts, while desperately well intended, are costly to all concerned and not especially effective in enhancing survival and reproductive fitness. We must therefore again ask: what, then, can be done?

A main part of the answer lies with confronting death anxieties and understanding deeply how humans deal with these fears—and the costs involved. If we face rather than deny personal death, we can deal with secured-frame anxieties and the vast negative consequences of frame deviations. Denial of death leads to denial of the importance of reality in emotional life, and with that comes a denial of the nature and functions of rules, frames, and boundaries. Denial begets denial, and denied realities cannot lead to effective adaptive responses or to any kind of learning and advancement in adaptive resources—denial is a dead end. Neither life in general nor psychotherapy in particular can advance in quality and healing as long as there is so much denial in place. Of course, we all need a measure of denial, but its pervasive use has blinded us to many critical realities and rendered us unable to advance our field and our lives.

As Plotkin (1994) and others have pointed out, we are not limited to automatic, genetically programmed intelligence. When it comes to emotional adaptation, it is imperative that we part ways with natural selection for a while. We need to trigger-decode our own messages and those from others so that we can overcome our lack of a natural, in-built decoding capacity; this is, for now, the only way we can consciously experience emotional reality to its fullest. And we need to forego denial through frame breaks in order to study scientifically the nature of frames, so that we can come to seek and accept the secured frames we so badly need yet so terribly dread.

We are experiments in evolutionary history, but we have evolved the faculties with which to be less than passive experimental subjects and become the active masters of our fates. But we need to stop deluding ourselves and take on these issues full-force—we are dealing with incredibly strong selection pressures and design flaws that can be ameliorated only through the proper use of the incredibly gifted intelligence that nature has given us—including its self-organizing emergent features. It is time we used this gift to our distinct advantage.

Finally, we need to fathom deeply the means by which nature creates variants of emotion-processing minds and how natural selection tests these variants and selects the best of them for favoured reproduction. As hominids have been making advances culturally and technically, the forces and rules of evolution appear to be changing. We need to probe deeply into these changes and come up with a new and contemporary theory of evolution, especially as it applies to the great mystery of how the emotion-processing mind has evolved in the past and will evolve in the future. In solving the mysteries of evolution, our prospects for favoured evolution and continued existence will greatly improve—for each of us personally, and for all of *Homo sapiens sapiens*.

As a step in this direction, I conclude this book with two chapters that will take us into a new area of exploration that evolutionary principles have given us—*mental Darwinism*.

PART IV

THE EMOTION-PROCESSING MIND AS A DARWIN MACHINE

CHAPTER FIFTEEN

Mental Darwinism

Having mapped a vast territory related to the evolution of the emotion-processing mind, in these final two chapters of the book I offer some new ideas on the current adaptations of this mental module and the nature of its operations. The central propositions for this discussion arise from the broad applicability of the Darwinian principles of evolution to adaptive entities and their functioning in immediate situations. The quest here is to determine and theorize about the extent to which these principles apply to the operations of the emotion-processing mind.

The central question, which can be asked in a variety of ways, is this: is the emotion-processing mind a so-called Darwin machine (Plotkin, 1994)? That is, is this mental module an entity—a basic human adaptive processing structure—that operates according to the Darwinian principles of evolution; does it function according to the rules of universal Darwinism (Dawkins, 1983)? Is it an entity that adapts and copes in a manner similar to other cognitive and language structures by means of the principle of *selectionism* (as opposed to the principles of *instructionism*; Gazzaniga, 1992)—or does it operate via a mixture of both principles? In still other terms, given the evidence collected by Edelman (1987, 1992) for *neural Darwinism*—is there also a *mental Darwinism*?

As I try to show, the answer to these equivalent questions appears to have profound consequences for both psychoanalytic theory and practice.

The basic thesis

To turn these questions into a positive statement, this chapter explores the central proposition that the emotion-processing mind has not only evolved over time in accordance with the Darwinian principles of evolution, but, as a cognitive module of the mind, it operates and adapts on a day-to-day basis in significant ways according to the same evolutionary principles. This thesis is then further refined to propose that the overall emotion-processing mind and its deep unconscious system function almost entirely according to selectionist rules and principles in which the environment chooses from largely in-built, available inner resources. On the other hand, the conscious system of the emotion-processing mind operates according to a superficial form of instructionism, through which the environment directs the operations of the system, and a deeper form of selectionism.

An initial perspective

To clarify these basic postulates, they imply that the same principles that prevail for competition *between* organisms apply *within* the human organism as well—including the functioning of the emotion-processing mind. That is, Darwinian principles operate both inter-organismically and intra-organismically; the same forces and rules that differentiate the propagation of organisms also differentially propagate their inner, adaptive resources. Thus, the principles related to the long-term development of competing individuals and their species apply equally well to aspects of their immediate functioning. Both processes adhere to the constraints of variation→test and selection→differential reproduction→resultant adaptation→introduction of fresh variants→repetitions of this cycle.

These principles operate as a set of rules and constraints that create the essential principles of *mental Darwinism*. They are:

1. The existence in humans of genetically determined, innate structures or entities based largely on chance variants. Because these structures are inherited and therefore screened through the Darwinian filter of natural selection, these entities exist because their efficiencies have led them to be selected during previous test periods.
2. Competition between existing variants for adaptive success.
3. Selection, by environmental events, for favourable reproduction of those variants that are most successful in terms of adaptation and survival.
    a. As a supplement to these broadly accepted principles, there is an additional selection factor in the ability of existing, already selected inner resources to evolve internally by *self-organizing and restructuring* to create emergent adaptive resources not previously available to the organism. This includes the development of unprecedented, creative adaptive solutions to existing and future environmental impingements (Kauffman, 1995).
4. A fresh set of chance and selected variants, which renew this evolutionary cycle.

The central theorem of *mental Darwinism* is that the emotion-processing mind possesses in-built, genetically determined attributes and adaptive capacities—it is not a *tabula rasa*. Environmental events select from these capacities the most optimal responses to their impingements and cause the preferential reproduction of these favourable adaptive reactions. The theory also allows for creative internal responses to environmental stimuli through the selection of innovative combinations of internal resources and the self-organizing emergence of unprecedented combinations of these resources.

Adaptation

There are two basic ways that adaptation can and does take place—*via instruction and via selection*. There are both pure forms and intermixtures of these two modes of operating, and some degree of intermingling is the rule.

Instruction

Instruction is essentially a Lamarckian theory that states that an organism's learning and adapting occurs strictly through instructions and directives from the environment. The organism is a *tabula rasa* with vague, if any, potentials, and the environment directs the organization of these potentials into an adaptive response. This theory affords to the environment the total power over learning and adaptive responsiveness; the organism cannot do anything beyond that which it is directed to do by its impingements and settings. Creative responses derived from within the organism are all but impossible—there is no basis for inner innovation. Furthermore, as learned responses, these instructed adaptations may serve socially as models for future generations, but they cannot be passed on through genes. Even though the theory is Lamarckian, it acknowledges that genes operate down a one-way street that runs from the genetic configuration to the mind and body, and not the reverse.

With regard to psychotherapy, instructionism would imply that a therapist could in various ways instruct a patient to change, and that under the right conditions this change would transpire as directed. With some evident modicum of success, instructionistic interventions are in common use and underlie almost all of today's psychotherapy practices, including psychoanalysis. However, the instructionist model is incomplete and does not account for the complexities of the emotion-processing mind. The following are of note:

1. Instructionism fails to appreciate that a therapist cannot directly instruct those aspects of conscious functioning that are under the sway of deep unconscious influence.
2. The model also fails to understand that a therapist cannot

instruct the very powerful deep unconscious part of the patient's mind, which is inaccessible to direct communications and messages.
3. A therapist's use of instructionism is highly biased in that, in intervening, the therapist fails to select from the patient's meaningful communications and instead imposes his or her own formulations on the patient—one reason why instructionism is so popular.
    a. Therapists who do not trigger-decode, who fail to grasp the *unconscious* meanings contained in their interventions, and whose instructionally oriented interpretative or directive focus imposes itself on, rather than selects from, the patient's own resources, do not make use therapeutically of their patients' vast conscious and unconscious inner capabilities.
4. Instructionism also involves a discrepancy between the *consciously intended* meanings of a therapist's interpretations and other interventions, and the meanings that a patient *deeply unconsciously perceives and processes*. Whenever a therapist tries to instruct a patient on how to deal with a particular situation, his or her deep unconscious system will be focused on and will be processing the therapist's need to direct, instruct, and dominate the patient—rather than on the consciously acknowledged meanings of the intervention itself.

Although it entirely neglects deep structure, instructionism is a model that may be applied to a limited extent and superficially to *conscious-system* forms of psychotherapy, where the effort is made to deal with the manifest contents–conscious thinking of a patient without regard for unconscious experience and forces. Instruction appears to be a factor in such efforts as deconditioning, cognitive therapy, and all forms of intervening carried out in terms of the implications of a patient's surface behaviours and associations—including all types of clarifications, confrontations, directives, advice, and the like.

There is, however, an evident weakness in instructionistic effects, as shown by the over three hundred forms of instructionistic psychotherapy. Indeed, as noted, these therapies do not allow the patient to make use of his or her own in-born and personally developed creativity and resources, and they also disregard the power of the deep unconscious system of the emotion-processing mind; they imply that a therapist can, in pygmalion-like fashion, create a patient's persona. Whatever directives can accomplish, they cannot influence the more powerful selectionism that is operating deeply and basically within the emotion-processing mind, and which actually may underlie whatever seeming success instructionist practices may have.

Selection

Selection, which is at the heart of Darwinian theory, operates very differently (for details, see Gazzaniga, 1992; Plotkin, 1994). As noted, the theory of selectionism has been successfully applied to the basic adaptive systems of *Homo sapiens sapiens*—the brain, language, and immune system and, quite broadly, to cognitive mental capacities, including the use of intelligence. Selectionism is a modern-day theory of considerable explanatory power, and it needs to be applied to psychoanalysis and psychotherapy and to the emotion-processing mind.

As a way of giving substance to this theory, the following is an expanded version of the main postulates and principles of selectionism:

1. An organism is born with genetically determined resources, some fixed and others flexible. Some of these are designed through evolution based on earlier experiences with the environment; others are random variants. There are no blank slates simply awaiting instructions from without; organisms are well prepared to adapt.
2. The organism's adaptive resources exist as potential capabilities that are affected by developmental processes under the influence of environment constraints.
3. Environmental impingements select from an organism's inner potentials those capabilities and configurations that produce the most optimal adaptive responses to external, adaptation-evoking stimuli.

4. These favourable responses are then differentially reproduced. This means that, mentally, they are repeated as further environmental impingements arise.

   a. The design of the emotion-processing mind is such that any form of partial repetition of the original, selecting stimulus evokes a repetition of the selected response—even when it is maladaptive.

5. There are, however, additional random variations generated by the flexible human mind. The human mind is a *self-organizing entity*, and, therefore, based on its natural and learned resources, it is capable of reacting to incoming stimuli with emergent *novel and creative* adaptive responses. Selection mechanisms ensure that the mind is not enslaved to the environment and that it has a measure of autonomy from its impingements. Creative responsiveness and advances in adaptive capabilities are therefore feasible in connection with selection-governed coping resources, though not in those ruled by instruction.

6. There is, however, a critical constraint to selectionism of considerable importance to psychoanalysis and psychotherapy. In the therapeutic situation, we are, of course, concerned with a patient's existing adaptations and maladaptations and how they operate, but we are even more concerned with how to change deeply and constructively a patient's maladaptive behaviours and symptoms—that is, *in these terms, how to modify dysfunctional selections* (see below).

In instructionism, change is brought about through directives. But in selectionism, change is brought about by a patient's abandoning existing maladaptive variants and shifting to new variants in their place. That is, a Darwin machine *does not have the capability through evolutionary principles of altering an existing adaptive response*; it can only allow that response to become extinct or die away and turn to—select—a more favourable alternative in its place.

This realization opens the door to an entirely new approach to the question of cure in psychotherapy and psychoanalysis. It compels us to take a fresh look at how therapists can effect psychic change and create lasting constructive modifications of maladaptive behaviours and symptoms. And it takes us to the deepest possible level (at present) at which such change can take place, because we are seeking to generate changes in the configuration of the emotion-processing mind—the fundamental adaptive module in the emotional domain.

> a. The issue of how to enable an organism to select new and innovative adaptive responses—and, especially, how to enable it to extinguish maladaptive selections in favour of those that are more effective—is a major unanswered question in evolutionary psychoanalysis.

Inborn resources

Lacking a strong evolutionary and genetic position, Freud (1923b) at first saw the id, defined in terms of sexual and aggressive instinctual drives, as the inborn, genetically determined aspect of the psyche and as the source from which the ego and superego were developed. Later analysts, like Hartmann (1939), with Kris and Lowenstein (Hartmann et al., 1951), argued that the ego has inborn features such as genetically determined capacities to relate, test reality, learn, adapt, and the like. On the other hand, the superego was seen as possibly having a biological core, but its structure was viewed mainly as a consequence of the internalization of experiences with, and attitudes gleaned from, the parents—as more of a developmental than an inborn structure.

More recently, evolutionary psychoanalysts have argued for the importance of inborn, genetically determined features for the three systems posited by the structural hypothesis and by object relations theory—including so-called relational structures (Slavin & Kriegman, 1992) and psychic defences (Nesse, 1990b; Nesse & Lloyd, 1992). These biological givens constitute the inherited structures and cognitive capacities that afford humans their adaptive resources, and they have been conceptualized largely in terms of their antecedents in animal, insect, and other non-human biological organisms. While these structures are also seen as open to some degree of environmental influence, the greater stress has been on genetic make-up—a corrective for Freud's overemphasis on the intrapsychic consequences of external traumas.

This analytic literature draws heavily on the work of Tooby and Cosmides (1987, 1990a, 1990b, 1992), who have argued that cognitive capacities have been laid down genetically during the Pleistocene era and that current environmental events select from, but do not instruct, available genetically established cognitive capacities like learning, memory, adapting, and language. Gazzaniga (1992), whose efforts were concentrated on the human cognitive mind as a group of functionally specific modules that operate according to selection principles, has offered an extensive history of these developments (see also Tooby & Cosmides, 1992b, for their support of the cognitive module concept). Gazzaniga has developed a number of convincing arguments for the operations of the cognitive aspects of the human mind as a Darwin machine—a telling insight that is based on the work of Jerne (1955, 1967) with the immune system.

On the whole, then, evolutionary psychoanalysts have been striving to define the genetically founded psychological and relational givens on which development and the environment act. These writers have not to any notable extent considered the problem of instructionism versus selectionism, nor have they explored the adaptive functioning of the emotion-processing mind and its genetic heritage. Nevertheless, they have provided us with a foundation for the study of the adaptive operations of this mental module—ready for us to build on here.

The emotion-processing mind as a Darwin machine

Returning to the emotion-processing mind, the key proposition is that this adaptive entity is a Darwin machine that operates according to the rules of evolution—selection rather than instruction. It therefore is a domain-specific, species-specific, functionally specific module of the mind, programmed to know what it must learn and adapt to, and with the potential resources to do so. As mentioned above, even though the *conscious system* operates through a mixture of instructionism and selectionism, the latter is more powerful and critical to emotional adaptation. In any case, the overall operations of the emotion-processing mind and of its deep unconscious system are governed exclusively by selectionism.

There is considerable observational and theoretical support for the thesis that the emotion-processing mind functions as a Darwin machine. The following arguments are relevant:

1. All of the basic adaptive systems and cognitive modules in *Homo sapiens sapiens* appear to operate as Darwin machines. We would therefore expect this principle to hold for the emotion-processing mind—the fundamental cognitive module and adaptive system in the emotional realm. This argument is in keeping with the principle that nature tends to conserve its adaptive and evolutionary strategies. In addition, given that the emotion-processing mind utilizes language and many cognitive functions that themselves operate as Darwin machines, we can expect that the emotion-processing mind does so as well.

2. The emotion-processing mind is an entity and structure that processes in-coming information and meaning and brings two levels of intelligence to bear on these inputs, resulting in adaptive output responses. This pattern and the design that underlies this module are universal for *Homo sapiens sapiens*, though with the expected individual variations. As such, the operations of the emotion-processing mind must, of necessity, be based on a series of complex, genetically determined substructures, which are the foundation for selective responses to external stimuli.

3. The emotion-processing mind is capable of highly innovative and individually distinctive adaptive responses of the kind that are possible only with selectionism.

4. Clinically, the fixity of adaptive responses once they have been established, even when they are largely symptomatic and therefore maladaptive, also is in keeping with selectionism. Emotional dysfunctions are not easily, if at all, modified through direct instruction. This suggests that once a dys-

functional organismic response has been selected for, it is difficult to interrupt the preferred replication of a selected maladaptive choice in favour of one that is more adaptive. This point is in keeping with evolutionary observations that indicate that once a mode of adaptation is established or a species has been defined, both the adaptation and the species tend to remain relatively unchanged for hundreds of thousands of years (Eldredge, 1995). It is also consonant with the finding that evolutionary principles do not include ways of altering selected adaptations, but instead allow for change in survival strategies only through the extinction of a given set of adaptations and the selection of new and more effective ones.

   a. This fixity of a maladaptation in the face of instructions to surrender the dysfunctional adaptation suggests that the conscious mind operates in important ways through principles of selectionism—and supports the thesis that the emotion-processing mind as a whole does so as well.

   b. On the other hand, the success of processes like deconditioning a maladaptive response, such as a fear of flying or of heights, indicates that the conscious system also operates via instructionism—it is open to environmental dictates regarding a fresh adaptive response. Thus, the conscious system appears to operate with mixed principles and is amenable to therapies that use either or both principles to guide their therapeutic efforts—direct confrontation or advice versus whatever it takes to modify selected maladaptations deeply (see below). However, there is evidence that even though surface behaviours appear to be open to instruction, this is illusory or, at the very least, a weak effect, beneath which a stronger set of selectionistic principles prevails.

   c. The deep unconscious system of the mind and the overall basic architecture of the emotion-processing mind do not appear to be amenable to change via instruction. This speaks for the argument that their operations follow selection principles.

5. As noted, overall, the design and operations of the emotion-processing mind appear to be selectionistically fashioned

and reinforced in that they are exceedingly difficult to alter through our current, essentially intructionistic, ways of doing psychotherapy.

    a. Despite intense directives from the environment to behave and operate otherwise, the defensive alignment of the conscious system of the emotion-processing mind is relatively fixed in its antagonism both to trigger-decoding and to bringing into awareness deep unconscious experiences, meanings, and insights—despite their great adaptive value.

        i. This is an example of a selected design feature of the emotion-processing mind that appears to be extremely difficult, if not impossible, to change by any presently known instructionistic or other means. This problem creates a distinct need to discover the techniques through which we can alter selected, maladaptive design features of the emotion-processing mind using insights drawn from the principles of natural selection.

    b. Similarly, once the emotion-processing mind has been damaged through environmental trauma and reconfigured to utilize overwhelmingly strong conscious-system defences that block virtually all access to deep unconscious system processing (Langs, 1995a), it is exceedingly difficult to reconfigure so that it becomes more efficient, non-defensive, successfully adaptive, and less self-harmful in its workings. Here, too, instructionism fails to set matters right, another indication that selectionism is at work.

Innate design features

There are strong reasons, then, to appreciate the implications of the proposition that the emotion-processing mind is a Darwin machine. Given the stress on the genetic foundation of this mental model, it behoves us to take a more detailed look at its innate features. On this basis, we may derive some clues as to how those features that are most maladaptive can be altered at any point in the course of a lifetime.

There is a universal design to the structure of the emotion-processing mind, yet each individual module is developed and configured in singular fashion through a combination of genetic, developmental, and experiential factors (Gazzaniga, 1992; Plotkin, 1994). The following aspects of the architecture of the emotion-processing mind seem pertinent:

1. There is a perceptual selection mechanism through which the implications and meanings of incoming stimuli are rapidly assessed and a threshold established so that certain meanings are directed towards the conscious system and conscious registration, while other meanings and their potentially disruptive qualities are sent to the deep unconscious system. There is, then, an *unconscious message-analysing centre* that translates incoming language sounds into symbols with meanings, and then sorts out the emergent meanings and assigns them for processing to either of the two systems of the emotion-processing mind.

    Because this gating/directing mechanism involves meaning, it undoubtedly is language-based. Therefore, it probably functions in infants in rudimentary fashion until language and verbalizable meaning, including the understanding and use of symbols, are well established. It is likely, too, that there are genetically programmed changes in how infants process emotionally charged experiences, so that around age 3 years, there is a shift from affect directed to meaning-directed processing, and from the use of global to quite specific processes.

    a. This gating mechanism and its thresholds appear to be under the influence of environmental events, and it is therefore a subsystem of the emotion-processing mind that is open to selection pressures. In a sense, then, given the many possible threshold and meaning-sensitive configurations built into this mechanism, significant environmental impingements selectively affect its operation.

b. In general, this subsystem appears to have its own developmental history, yet once its design is set, it appears to be especially vulnerable to major death-related traumas, which increase its threshold for the conscious registration of emotionally charged meanings. *The very rules by which selectionism operates may therefore be affected by environmental events.*

c. In addition, with ageing, there may be natural or universal changes in this screening device—especially in later decades when personal mortality moves ever closer to actualization.

2. Both conscious and unconscious perception operate via all five senses, although vision and audition seem most prominent for the functioning of the emotion-processing mind.

    a. The evolutionary course and individual development of unconscious perception is a scientific study in need of further pursuit. As the basis for cognitive functioning, emotional and otherwise, perceptual systems tend to be relatively stable and smooth-functioning, although they also appear to be open to selection mechanisms and are vulnerable to damage or reconfiguration.

        i. There is an evident trend for environmental trauma, especially when it is severe or death-related, to affect and fix the selection of maladaptive configurations in the emotion-processing mind.

3. Conscious-system processing—perception, knowledge, intelligence, memory storage, adaptation—involves subsystems that are based on genetically founded features. These attributes of the conscious system are made up of actual and potential repertoires of resources that are selected by external events and, secondarily, by internal reactions—conscious and unconscious—to these events.

    a. There is an extraordinary degree of complexity to these vital subsystems, each of them hard-wired at bottom, but flexibly designed as well—a feature that allows for inventiveness and creative responsiveness, and for the development of pliable coping capabilities. It seems, then, that selectionism implies both choice by environmental impingements and choice by the organism—a critical evolved feature of selectionism that is most advanced in *Homo sapiens sapiens* and allows for maximal autonomy from the environment.

b. A notable and relatively fixed feature of conscious-system processing is its capacity to take in and adapt to a very wide range of environmental inputs, although there are thresholds that limit the human range of sensory and meaning sensibility. In addition, however, the system shows an in-built, relative imperviousness to excessively disturbing emotionally charged meanings and to frame impingements, and it has a general preference for frame deviations. The relative immutability of the conscious system with regard to frame-related responses and choices speaks for genetic wiring and selectionism.

c. We are in a position to understand a great deal about the selected design of the emotion-processing mind and the events that can lead to maladaptive selections. But while we know in general that favourable and positive life experiences in childhood and to some extent in later life, including a sound psychotherapy experience, may lead to constructive adaptive selections, the means by which dysfunctional selections can be replaced with functional selections is far less clear.

4. The deep unconscious system is another processing system that has a basic genetic foundation that provides it with its distinctive features—for example, capabilities for processing multiple bits of information and meaning simultaneously, and intense frame sensitivity with a genetically grounded appreciation for secured frames. The deep unconscious system operates with its own intelligence, which appears to be highly adaptive in its choices. All of these capacities have innate foundations and, therefore, are affected by selectionistic principles.

5. Finally, I have previously alluded to aspects of the wiring or tracts of the emotion-processing mind. Here, too, we are dealing with a highly stable universal configuration, individually crafted.

a. Most critical in this respect is the absence of mental tracts from the deep unconscious wisdom subsystem to the conscious system, thereby depriving humans of access to and use of their most effective knowledge of the emotional domain. Instruction has virtually no effect on this evolved and selected arrangement, which is reinforced through selectionism.

b. Much the same applies to the fear–guilt subsystem and its power over conscious-system choices and behaviours.

c. The emotion-processing mind appears to have been selected for a strong measure of self-harm in all humans. Clearly, we need to develop ways of repairing the damage caused by natural selection. It is as if we are all born with a congenital disease like thalassemia (a blood dyscrasia). This illness has been favourably selected because the genes involved also confer a relative immunity to malaria. Similarly, we are all born with a congenitally deficient emotion-processing mind, which confers a relative immunity to disruptive unconscious meanings while conferring on us the inherited disease of a dysfunctional emotional-processing mind (see also Nesse & Williams, 1994).

d. Strange indeed is the argument that the emotion-processing mind is inherently a relatively dysfunctional selected system. This attribute makes the modification of its selected capacities enormously critical.

6. Finally, we may note that each of the systems and subsystems of the emotion-processing mind is based on complex genetic arrangements that are affected by developmental and environmental factors. As with the human brain, there is an enormous amount of mental adaptive machinery for the environment to select from, and select it evidently does. How to help it select for better ways of living and coping, and to modify its maladaptive choices, are among the greatest challenges for today's psychoanalysts and psychotherapists.

CHAPTER SIXTEEN

Selectionism and the therapeutic process

The theory of selectionism appears to challenge and recast a number of clinical issues related to both psychotherapy and psychoanalysis. In this last chapter, I attempt to develop some clinical implications of this new perspective on emotional adaptation and its dysfunctions. My main focus will be on the ramifications of the aforementioned concept that, in large measure, the emotion-processing mind operates as a Darwin machine.

We can begin by identifying the key clinical dilemma posed by this theory. Stating that the emotion-processing mind adapts and functions according to the principles of universal Darwinism and selectionism implies that emotional dysfunctions are a consequence of maladaptive selections made by environmental events with respect to the repertoire of resources within the emotion-processing mind. In essence, then, psychopathology, most broadly defined, is at root a consequence of selections of maladaptive strategies for the operations of the emotion-processing mind.

As we have seen, the central question that follows from this formulation is this: how can a psychotherapist help a patient to modify effectively his or her unconsciously driven use of pathological or dysfunctional, maladaptive selections and replace them in some lasting way with relatively effective adaptive processes and responses? How do we "cure" the selected maladaptations of the emotion-processing mind?

The nature of maladaptive selections

Let us begin this pursuit by defining the sources of maladaptive selection processes. There appear to be two basic types of maladaptive selections open to possible modification in psychotherapy:

1. *Evolutionary-based selections.* This refers to our inherited genetic make-up as it determines the basic configuration of the emotion-processing mind. As we have seen, this evolved configuration is in many ways structurally maladaptive, thereby raising the problem of how we can alter these genetically selected attributes of the emotion-processing mind in order to improve an individual's adaptive repertoire and emotional life.
2. *Experientially and environmentally guided selections.* This refers to critical life experiences, favourable–gratifying and unfavourable–traumatic, that affect:
   a. the basic structure of the emotion-processing mind and its fundamental processing methods;
   b. specific aspects of the processing of emotionally charged information and meaning—specific adaptive techniques or survival strategies.

Selection theory and clinical practice

The issue we must consider now revolves around the means by which each of these dysfunctional forms of selection are made and sustained, and how they can be modified in psychotherapy.

To offer some tentative answers to these unfamiliar and difficult questions, I present some suggestions as to how selection theory affects clinical practice:

1. Selection theory explicitly states that adaptations are interactional and that environmental impingements are the primary cause of chosen responses. This implies that emotionally related adaptations, successful and unsuccessful, are triggered and selected by external events. As noted earlier, this postulate speaks against that part of psychoanalytic theory that views inner causes—fantasy formations, memories, and the like—as *basic or primary* causal factors in neurosis.
   a. In emotional maladaptations, the *primary* causal factor is environmental, but once a selected response has been established and favoured, it may *secondarily* become an autonomous cause of emotional dysfunction. Thus, dysfunctions are always the result of an interaction between an individual's genetic endowment, his or her environmental impingements, and the nature of sustained inner responses, conscious and unconscious. Selection processes play a role in each of these factors.
      i. The essential sequence is: genetic make up →environmental event → selected maladaptive mental response → symptomatic effects → favoured reproduction of the response → and the generation of new variants to prepare for further compelling external events.
   b. This line of thought leads to a definition of neuroses (emotional dysfunctions) as maladaptations that are selected by environmental events and then afforded preferential reproduction—used again and again.
   c. Broadly speaking, learning theory proposes that only gratifying and positive or effective adaptations should find reward, reinforcement, and favoured reproduction. This rule prevails in general in nature, yet with *Homo sapiens sapiens*, and especially in the emotional realm, often this simply is not the case. Maladaptations that are negatively toned and harmful to self and to others continue, once selected, to be favoured.

i. We need to know a great deal more about the unconscious factors involved in emotionally laden learning and reinforcement. Somewhere in understanding why maladaptive solutions are so staunchly maintained lies the beginning of the answer to how to make it otherwise—that is, how to alter maladaptive selections or emotional learning dysfunctions. In this regard, ideas about the repetition compulsion as a belated attempt at mastery or a form of masochism (Freud, 1920g) are overly simplistic and do not solve this problem.

d. One aspect of the learning process that seems to misfire in emotional learning is *over-generalization*. That is, once a trauma has been suffered and a maladaptive response has been selected, every event that is even remotely similar to the original trauma is likely to evoke a repetition of the selected response, even when it is dysfunctional. Here too, nature seems to have opted for overprotection in a way that leads to emotional suffering.

e. A second dysfunctional aspect of the emotional learning process involves the intense fixity of selected maladaptive responses. While the reproduction of learned adaptive responses that are effective and beneficial obviously favours survival, replicating maladaptive responses diminishes flexibility and creativity and seems to diminish chances of survival. There appears to be a lack of discrimination in the selection and replication processes, with a failure to reinforce adaptive rather than maladaptive responses differentially—"once selected, ever fixed" seems to be the principle, regardless of the adaptive success of the response.

f. It seems likely that psychotic states arise under one or two possible conditions:
   i. In-born, genetically caused dysfunctions of the emotion-processing mind—in which case, the repertoire from which the environment can select is a restricted and largely dysfunctional one.
   ii. Severely traumatic environmental events, acutely or

chronically damaging to the emotion-processing mind, that forcibly select psychotic adaptations.

As a rule, contributions from both sources are to be expected.

2. This work raises profound questions about existing theories and research on the process and nature of therapeutic cure. The Freudian and post-Freudian theories of psychoanalysis on which the ideas about symptom alleviation are based are essentially instructionistic theories of change. Selectionism casts doubts on the validity of these ideas and calls for an entirely different approach to this problem, one that includes a selectionistic theory of cure. Such an approach would be consonant with evolutionary principles and the findings of evolutionary psychoanalysis. After all, the latter is a more fundamental subscience of psychoanalysis than psychodynamics, and therefore it should offer a more fundamental and compelling basis for the investigation of the therapeutic process of deep psychological change and cure.

   a. Current theories of psychic change, whether focused on the ego, id, and/or superego or on relationship structures, tend to be unsupported hermeneutic fictions. Communicative ideas about trigger-decoded insights, positive unconscious introjective identifications with well-functioning therapists, and about the healing powers of secured frames, have more substance, but they, too, have failed to capture the essence of the curative process.

   b. In general, the concept of dysfunctional emotional symptoms and interactions—of *psychopathology*—is used in a manner similar to the way the term *fever* is used as it pertains to physical illness. Psychological cure as it is now defined is analogous to a patient's simple report to a physician that his or her fever has gone (most outcome studies are based on questionnaire responses). We are lacking a measure of "psychic temperature" and a way of taking it in different regions of the mind (cure in one area, illness in another). But even if we had such measures, a fever is merely a surface manifestation of an underlying illness that is, itself, quite specific in nature. A fever may be caused by many different kinds of dysfunctions in many different systems of the body. It also arises because of the ways in which each of these local dysfunctions affects the brain's temperature-control mechanisms.

There is a hierarchy here, with the fever at the highest level, organ disease at the next, and disturbances in temperature regulation at the core of the matter. Similarly, we need to conceptualize illness and cure in the emotional domain in a hierarchical manner. Emotional symptoms would be closest to the surface, specific mental-system dysfunctions next, and disturbances in the control and processing functions of the emotion-processing mind would be at the core. We are in need of measures of function and dysfunction, adaptation and maladaptation, at each level of this critical hierarchical constellation.

    i. We should take a jaundiced view of present theories of cure because the encoded meanings of the material from patients in psychotherapy and the fundamentals of the emotion-processing mind have gone virtually unrecognized by today's psychotherapists. Indeed, the entire realm of deep unconscious experience has been neglected.

    ii. A sound theory of the process of cure is likely to be fashioned from a sophisticated and deep understanding of selection theory and of the nature of psychopathology, especially at its most fundamental level—errant selections made by the environment with respect to the emotion-processing mind.

c. As it pertains to adaptive dysfunctions in patients, the selection process operates on at least two levels:

    i. The surface level of the *symptomatic disorder*, which includes the full range of psychologically based maladaptations—emotional symptoms such as phobias and obsessions, characterological disorders, interpersonal disturbances, psychosomatic dysfunctions, and the like. On this level, the goal of therapy is to enable the patient to discard his or her symptomatic selection choices in favour of those that are non-symptomatic. In this regard, it is well to realize that Darwinian principles play a role in the perpetuation of symptomatic disorders.

    ii. The level of *basic intrapsychic functioning*, the dysfunction underlying the symptomatic picture. On this level, we are dealing with disorders of processing and adapting to emotionally charged information and meaning—*with dysfunctions of the emotion-processing mind* (Langs, 1995a). Here, selection has opted to choose maladaptive forms of processing that need to be surrendered in favour of more optimal ways of adapting to incoming impingements.

d. In this model, the external world receives full consideration as the so-called *fitness environment*, the reality situations that select from the repertoires of internal capabilities within the structure of the emotion-processing mind. Clearly, relatively supportive or benign environments will select very differently from those environments that are traumatic in aggressive, seductive, and other ways. Similarly, frame-secured environments select very differently from those that are essentially frame-deviant.

    i. A challenge for cure through psychotherapy arises when individuals continue to seek out traumatic environments and relationships even though they have freed themselves from the hurtful situations within which selections for symptoms and dysfunctional processing methods took place.

    ii. A relatively unrecognized problem exists for psychotherapists whose frame-deviant environments and incomplete or erroneous interventions (which are not based on an adaptive viewpoint and trigger-decoding) actually unwittingly create situations that are, for the patient (and themselves), hostile, seductive, frame-deviant, and otherwise traumatic. Under these conditions, the therapy fitness environment actually replicates the patient's earlier pathology-evoking environment and thereby unconsciously reinforces the mal-adaptive selections made by the patient in the earlier situations. It seems likely that under these conditions, no amount of instruction or newly discovered means of modifying maladaptive selections will change a patient's basic adaptive choices, because unconsciously they are being reinforced by the conditions of the therapy.

Selected maladaptations

Let us now consider some specific problems related to the curative process raised by a selectionist theory of the emotion-processing mind. Given the newness of these concepts, the ideas I present are rather speculative and tentative—heuristic probes far more than final answers.

First, a perspective: if we look at neuroscience for a moment, we discover that the theory of neural Darwinism has only recently been articulated and that all of the initial research has been directed at the many basic issues that arise from viewing the human brain as a Darwin machine. Problems of brain dysfunction are barely discussed from this new perspective.

In psychoanalysis, however, we seldom allow time for the study of normal mental functions before we demand ideas about dysfunctions and cure. Because we are fundamentally a healing endeavour belatedly seeking a science, rather than a science with a healing component, these demands are understandable—even if they are premature. Thus, the best I can do for the moment is to define some problems and offer a few ideas as to the directions from which their solutions may come. If I can establish a new way of looking at our clinical dilemmas and theoretical uncertainties, I will, I believe, have accomplished my main purpose at this early stage of investigation.

The following seem pertinent:

1. There are two types of selection processes at issue in this regard:
    a. *The selection process that guides the Darwinian evolution of the emotion-processing mind.* The clinical issues that arise in this regard are related to how the emotion-processing mind operates when it is intact; the causes of

its long-term and short-term dysfunctions; the exact nature of these dysfunctions; how they manifest themselves on the surface and in the depths; and how these dysfunctions can be modified, repaired, or cured (see Langs, 1995a, for an initial discussion of these issues).

b. *The selection process that pertains to the environmental directives to the emotion-processing mind to adopt particular strategies of adaptation, especially when these prove to be maladaptive.* Here, the clinical problem centres around finding ways of rectifying maladaptive patterns of response to environmental impingements, which, in selection theory, translates into their extinction and the generation of fresh solutions.

2. The dysfunctions caused by both of these selection processes may be termed *syndromes of pathological selection processes or of dysfunctional mental Darwin machines.*

3. Revisions of the dysfunctional aspects of the basic structure of the emotion-processing mind can be expected to be extremely difficult to achieve. The best solution I can presently envision involves ways of bypassing or altering the genetic design of the emotion-processing mind, mainly by finding definitive ways to enable the conscious system not only to tolerate access to deep unconscious meaning, but actively to seek it out and use it. This effort must combat a deeply established natural design configuration; we need new forms of therapeutic intervention to make it successful.

4. Modifications of actively selected maladaptive processing propensities calls for a new approach to psychotherapy that builds on ways of altering selected modes of mental response.

   a. In this regard, recall that selection processes do not have the means of modifying an existing, selected adaptive response no matter how maladaptive it may be. Once selected, it will be reproduced and repeated in situations similar to the original traumatic event until the means are found to extinguish it.

   b. Natural selection operates by allowing selections that prove dysfunctional to die out slowly, albeit unchanged. With the emotion-processing mind, this process either does not take place or transpires at an inordinately slow pace. We therefore need to find the means of facilitating and speeding up this change process or to find some other process that will do the job.

5. Another way of stating a central problem in this area is that the emotion-processing mind has been designed to operate smoothly under conditions of relatively mild to moderate emotional threat. However, when the level of threat and anxiety intensifies beyond a certain threshold (which is set at a particular level by each individual as a result of in-born and experiential factors), the emotion-processing mind tends to become overridingly defensive in its operations—conscious-system defences are intensified and staunchly maintained.

   a. In general, mild frame-securing and frame-deviating interventions by therapists evoke strong encoded derivatives and themes from patients that can, with some effort, be readily linked to their triggers for a successful decoding experience.

   b. On the other hand, intense frame-securing and strongly frame-deviant interventions lead to system dysfunctions in that trigger-decoding proves to be exceedingly difficult or impossible, despite every effort of the therapist. It is this type of selected processing dysfunction that leads to symptom formation and which is in need of modification through principles of therapy guided by Darwinian selection theory.

A final comment

Modifying selection processes has not been a pursuit of evolutionary biologists and psychoanalysts; they have been satisfied to allow evolution through natural selection to take its natural course. But the emotion-processing mind is very much another matter. As psychotherapists and psychoanalysts, we are dealing with one of the most unique and evidently incomplete and costly entities to have evolved in all of nature.

The stakes are high. The present exploration began as a quest for the clarification of psychoanalytic theory and for better ways of doing psychotherapy. But it has unearthed a dysfunctional mental system that appears to be placing our entire species at risk. Clearly, the very fact that we are threatening ourselves with extinction, and that we are unable to control our violent impulses towards each other reasonably, speaks for a dysfunctional mental design with serious consequences. Indeed, the realization that the emotion-processing mind is at the heart of civilization's current woes places psychoanalysis at the very centre of our struggles for personal and collective survival.

As psychoanalysts, we have a moral, ethical, and scientific obligation to establish the means by which we can explore the emotional basis for this crisis of survival. This means providing psychoanalysis with a sound and strong biological orientation, developing a formal science for the field, creating a viable and validated hierarchical theory with powers of prediction and change, making full use of the sub-theories of the field including evolutionary psychoanalysis, supporting the scientific pursuit of an improved design of the emotion-processing mind, and using the insights so gained to forge more effective modes of therapy.

Above all, we must use the fresh knowledge that we garner in these ways to help our species live in greater peace and emotional stability for centuries to come. There is, I believe no greater challenge facing psychoanalysis. The least measure of success will give our field the respect and rightful position in healing and science it has long striven for but has found all too elusive. We have a lot to overcome, but the rewards, personally and collectively, are almost beyond the imagination of our gifted and unfinished emotion-processing minds.

REFERENCES

Abbott, E. (1884). *Flatland: A Romance in Many Dimensions.* Oxford: Blackwell, 1962.
Alexander, R. (1979). *Darwinism and Human Affairs.* Seattle, WA: University of Washington Press.
Alexander, R. (1989). The evolution of the human psyche. In: P. Mellars & C. Stringer (Eds.), *The Human Revolution.* Princeton, NJ: Princeton University Press.
Badcock, C. (1986). *The Problem of Altruism: Freudian-Darwinian Solutions.* London: Basil Blackwell.
Badcock, C. (1990a). Is the oedipus complex a Darwinian adaptation? *Journal of the American Academy of Psychoanalysis, 18*: 368–377.
Badcock, C. (1990b). *Oedipus in Evolution.* London: Basil Blackwell.
Badcock, C. (1994). *PsychoDarwinism.* London: HarperCollins.
Baldwin, J. (1896). A new factor in evolution. *American Naturalist, 30*: 441–451, 536–553.
Barkow, J., Cosmides, L., & Tooby, J. (Eds) (1992). *The Adapted Mind.* New York: Oxford University Press.
Becker, E. (1973). *The Denial of Death.* New York: Free Press.
Bickerton, D. (1990). *Language and Species.* Chicago, IL: University of Chicago Press.

REFERENCES

Bickerton, D. (1995). *Language and Human Behavior.* Seattle, WA: University of Washington Press.
Bock, W. (1980). The definition and recognition of biological adaptations. *American Zoologist, 20*: 217–227.
Brown, D. (1991). *Human Universals.* New York: McGraw-Hill.
Brown, M. (1990). *The Search for Eve.* New York: Harper & Row.
Bruner, J. (1990). *Acts of Meaning.* Cambridge, MA: Harvard University Press.
Burnet, F. (1959). *The Clonal Selection Theory of Acquired Immunity.* Nashville, TN: Vanderbilt University Press.
Calvin, W. (1987). The brain as a Darwin machine. *Nature, 330*: 33–34.
Calvin, W. (1991). *The Ascent of Mind.* New York: Bantam.
Campbell, D. (1974). Evolutionary epistemology. In: P.A. Schilipp (Ed.), *The Philosophy of Karl Popper, Vol. 1* (pp. 413–463). La Salle, IL: Open Court.
Chomsky, N. (1980). *Rules and Representations.* New York: Columbia University Press.
Chomsky, N. (1988). *Language and Problems of Knowledge.* Cambridge, MA: MIT Press.
Corballis, C. (1991). *The Lopsided Ape.* New York: Oxford University Press.
Cosmides, L., & Tooby, J. (1992). Cognitive adaptations for social exchanges. In: J. Barkow, L. Cosmides, & J. Tooby (Eds), *The Adapted Mind* (pp. 163–228). New York: Oxford University Press.
Cronin, H. (1991). *The Ant and the Peacock.* New York: Cambridge University Press.
Darwin, C. (1859). *On the Origin of Species by means of Natural Selection.* London: Murray.
Darwin, C. (1872). *The Expression of the Emotions in Man and Animals.* London: Murray.
Dawkins, R. (1976a). Hierarchical organization: a candidate for ethology. In: P. Bateson & R. Hinde (Eds.), *Growing Points in Ethology.* Cambridge: Cambridge University Press.
Dawkins, R. (1976b). *The Selfish Gene.* New York: Oxford University Press.
Dawkins, R. (1983). Universal Darwinism. In: D. S. Bendall (Ed.), *Evolution from Molecules to Man* (pp. 403–425). Cambridge: Cambridge University Press.
Dawkins, R. (1987). *The Blind Watchmaker.* New York: W.W. Norton.
de Duve, C. (1995). *Vital Dust.* New York: Basic Books.

Dennett, D. (1995). *Darwin's Dangerous Idea.* New York: Simon & Schuster.
Donald, M. (1991). *Origins of the Modern Mind.* Cambridge, MA: Harvard University Press.
Edelman, G. (1987). *Neural Darwinism.* New York: Basic Books.
Edelman, G. (1992). *Bright Air, Brilliant Fire: On the Matter of the Mind.* New York: Basic books.
Eigen, M. (1992). *Steps Towards Life.* Oxford: Oxford University Press.
Eldredge, N. (1992). *Interactions: The Biological Context of Social Systems.* New York: Columbia University Press.
Eldredge, N. (1995). *Reinventing Darwin.* New York: Wiley.
Eldredge, N., & Grene, M. (1992). *Interactions: The Biological Context of Social Systems.* New York: Columbia University Press.
Eldredge, N., & Salthe, S. (1984). Hierarchy and evolution. *Oxford Surveys in Evolutionary Biology, 1:* 182–206.
Fagan, B. (1990). *The Journey from Eden.* London: Thames and Hudson.
Freud, S. (1900a). *The Interpretation of Dreams. Standard Edition, 4 & 5:* 1–627.
Freud, S. (1912–13). *Totem and Taboo. Standard Edition, 13:* 1–161.
Freud, S. (1915/1985). *A Phylogenetic Fantasy: Overview of the Transference Neuroses* (trans. A. Hoffer & P. Hoffer). Cambridge, MA: Harvard University Press, 1987.
Freud, S. (1915c). Instincts and their vicissitudes. *Standard Edition, 14:* 117–140.
Freud, S. (1920g). *Beyond the Pleasure Principle. Standard Edition, 19:* 3–64.
Freud, S. (1923b). *The Ego and the Id. Standard Edition, 19:* 1–66.
Freud, S. (1927). Postscript (to "The question of lay analysis"). *Standard Edition, 20:* 251–258.
Freud, S. (1950 [1895]). Project for a scientific psychology. *Standard Edition, 1:* 283–397.
Gazzaniga, M. (1992). *Nature's Mind.* New York: Basic Books.
Gill, M. (1994). *Psychoanalysis in Transition.* Hillsdale, NJ: The Analytic Press.
Gill, M., & Rapaport, D. (1959). The points of view and assumptions of metapsychology. *International Journal of Psycho-Analysis, 40:* 153–162.
Glantz, K., & Pearce, J. (1989). *Exiles from Eden.* New York: W.W. Norton.
Gould, S. (1980). *The Panda's Thumb.* New York: W.W. Norton.

Gould, S. (1982). The meaning of punctuated equilibrium and its role in validating a hierarchical approach to macroevolution. In: R. Milkman (Ed.), *Perspectives on Evolution*. Sunderland, MA: Sinauer.

Gould, S. (1987). Biological potentiality vs. biological determinism. In: S. Gould, *Ever Since Darwin* (pp. 251–259). New York: W.W. Norton.

Gould, S. (1989). *Wonderful Life: The Burgess Shale and the Nature of History*. New York: W.W. Norton.

Gould, S., & Lewontin, R. (1979). The spandrels of San Marco and the Panglossian paradigm. A critique of the adaptationist programme. *Proceedings of the Royal Society of London, 250*: 281–28.

Grene, M. (1987). Hierarchies in biology. *American Scientist, 75*: 504–510.

Griffin, D. (1984). *Animal Thinking*. Cambridge, MA: Harvard University Press.

Grossman, W. (1993). Hierarchies, boundaries, and representation in a Freudian model of Mental organization. In: A. Wilson & J. Gedo (Eds.), *Hierarchal Concepts in Psychoanalysis* (pp. 170–202). New York: The Guilford Press.

Hartmann, H. (1939). *Ego Psychology and the Problem of Adaptation*. New York: International Universities Press, 1958.

Hartmann, H., Kris, E., & Lowenstein, R. (1951). Some psychoanalytic comments on "culture and personality". *Papers on Psychoanalytic Psychology. Psychological Issues* (Monograph 14). New York: International Universities Press, 1958.

Jerne, N. (1955). The natural-selection theory of antibody formation. *Proceedings of the National Academy of Sciences of the United States, 41* (11): 849–857.

Jerne, N. (1967). Antibodies and learning: Selection versus instruction. In: G. G. Quarton, T. Melnechuck, & F. O. Schmitt (Eds.), *The Neurosciences: A Study Program, Vol. 1* (pp. 200–205). New York: Rockefeller University Press.

Johanson, D., & Shreeve, J. (1989). *Lucy's Child*. New York: Morrow.

Kauffman, S. (1995). *At Home in the Universe*. New York: Oxford University Press.

Kitcher, P. (1985). *Vaulting Ambition: Sociobiology and the Quest for Human Nature*. Cambridge, MA: MIT Press.

Kitcher, P. (1992). *Freud's Dream*. Cambridge, MA: MIT Press.

Kuhn, T. (1962). *The Structure of Scientific Revolution.* Chicago, IL: Chicago University Press.
Kuper, A. (1994). *The Chosen Primate.* Cambridge, MA: Harvard University Press.
Langs, R. (1982). *Psychotherapy: A Basic Text.* New York: Jason Aronson.
Langs, R. (1986). Clinical issues arising from a new model of the mind. *Contemporary Psychoanalysis, 22*: 418–444.
Langs, R. (1987a). A new model of the mind. *The Yearbook of Psychoanalysis and Psychotherapy 2*: 3–33.
Langs, R. (1987b). Clarifying a new model of the mind. *Contemporary Psychoanalysis, 23*: 162–180.
Langs, R. (1988). *A Primer of Psychotherapy.* New York: Gardner Press.
Langs, R. (1992a). *A Clinical Workbook for Psychotherapists.* London: Karnac Books.
Langs, R. (1992b). 1923: the advance that retreated from the architecture of the mind. *International Journal of Communicative Psychoanalysis and Psychotherapy 7*: 3–15.
Langs, R. (1992c). *Science, Systems and Psychoanalysis.* London: Karnac Books.
Langs, R. (1993a). *Empowered Psychotherapy.* London: Karnac Books.
Langs, R. (1993b). Psychoanalysis: narrative myth or narrative science? *Contemporary Psychoanalysis, 29*: 555–594.
Langs, R. (1994a). *Doing Supervision and Being Supervised.* London: Karnac Books.
Langs, R. (1994b). *The Dream Workbook.* Brooklyn, NY: Alliance.
Langs, R. (1995a). *Clinical Practice and the Architecture of the Mind.* London: Karnac Books.
Langs, R. (1995b). Psychoanalysis and the Science of Evolution. *American Journal of Psychotherapy, 49*: 47–58.
Langs, R. (1995c). *The Daydream Workbook.* Brooklyn, NY: Alliance.
Langs, R., & Badalamenti, A. (1992a). Some clinical consequences of a formal science for psychoanalysis and psychotherapy. *American Journal of Psychotherapy, 46*: 611–619.
Langs, R., & Badalamenti, A. (1992b). The three modes of the science of psychoanalysis. *American Journal of Psychotherapy, 46*: 163–182.
Langs, R., & Badalamenti, A. (1994a). A formal science for psychoanalysis. *British Journal of Psychotherapy, 11*: 92–104.

Langs, R., & Badalamenti, A. (1994b). Psychotherapy: the search for chaos, the discovery of determinism. *Australian and New Zealand Journal of Psychiatry, 28*: 68–81.

Langs, R., & Badalamenti, A. (in press). *The Cosmic Circle: The Unification of Mind, Matter and Energy.* Brooklyn, NY: Alliance.

Leakey, R., & Lewin, R. (1992). *Origins Reconsidered.* New York: Doubleday.

Lewontin, R. (1970). The units of selection. *Annual Review of Ecology and Systematics, 1*: 1–18.

Lewontin, R. (1979). Sociobiology as an adaptationist program. *Behavioral Science, 24*: 5–14.

Lewontin, R. (1983). On constraints and adaptation. *Behavioral and Brain Sciences, 4*: 244–245.

Liberman, P. (1991). *Uniquely Human.* Cambridge, MA: Harvard University Press.

Lloyd, A. (1990). Implications of an evolutionary metapsychology for clinical psychoanalysis. *Journal of the American Academy of Psychoanalysis, 18*: 286–306.

Mayr, E. (1974). Behavior programs and evolutionary strategies. *American Scientist, 62*: 650–659.

Mayr, E. (1983). How to carry out an adaptationist program. *American Naturalist, 121*: 324–334.

Miller, G. (1956). The magic number seven, plus or minus two. *Psychological Review, 63*: 81–97.

Millikan, R.G. (1984). *Language, Thought and Other Biological Categories.* New York: Bradford/MIT.

Millikan, R.G. (1993). *White Queen Psychology and Other Essays for Alice.* New York: Bradford/MIT.

Munz, P. (1985). *Our Knowledge of the Growth of Knowledge: Popper or Wittgenstein?* London: Routledge and Kegan Paul.

Nesse, R. (1990a). Evolutionary explanations of emotions. *Human Nature, 1*: 261–289.

Nesse, R. (1990b). The evolutionary functions of repression and the ego defenses. *Journal of the American Academy of Psychoanalysis, 18*: 260–285.

Nesse, R., & Lloyd, A. (1992). The evolution of psychodynamic mechanisms. In: J. Barkow, L. Cosmides, & J. Tooby (Eds.), *The Adapted Mind* (pp. 601–624). New York: Oxford University Press.

Nesse, R., & Williams, G. (1994). *Why We Get Sick: The New Science of Darwinian Medicine.* New York: Times Books.

Ornstein, R. (1991). *The Evolution of Consciousness.* New York: Prentice Hall.

Piaget, J. (1953). *The Origin of Intelligence in the Child.* London: Routledge & Kegan Paul.
Piaget, J. (1979). *Behaviour and Evolution.* London: Routledge & Kegan Paul.
Pinker, S. (1994). *The Language Instinct.* New York: Morrow.
Pinker, S., & Bloom, P. (1990). Natural language and natural selection. *Behavioral and Brain Sciences, 13*: 707–784.
Plotkin, H. (1990). *The Nature of Knowledge: Concerning Adaptations, Instinct and the Evolution of Intelligence.* London: Allen Lane/Penguin.
Plotkin, H. (1994). *Darwin Machines and the Nature of Knowledge.* Cambridge, MA: Harvard University Press.
Restak, R. (1994). *The Modular Brain.* New York: Schribner's.
Ridley, M. (1985). *The Problems of Evolution.* Oxford: Oxford University Press.
Ritvo, L. (1990). *Darwin's Influence on Freud.* New Haven, CT: Yale University Press.
Salthe, S. (1985). *Evolving Hierarchical Systems.* New York: Columbia University Press.
Shulman, D. (1990). The investigation of psychoanalytic theory by means of the experimental method. *International Journal of Psycho-Analysis, 71*: 487–497.
Simon, H. (1962). The architecture of complexity. *Proceedings of the American Philosophical Society, 106*: 467–482.
Simon, H. (1982). *The Sciences of the Artificial.* Cambridge, MA: MIT Press.
Slavin, M., & Kriegman, D. (1992). *The Adaptive Design of the Human Psyche.* New York: Guilford Press.
Smith, D. (1991). *Hidden Conversations: An Introduction to Communicative Psychoanalysis.* London: Tavistock/Routledge.
Smith, D. (1992). Where do we go from here? *International Journal of Communicative Psychoanalysis and Psychotherapy, 7* (1): 17–27.
Sulloway, F. (1979). *Freud: Biologist of the Mind.* New York: Basic Books.
Tooby, J., & Cosmides, L. (1987). From evolution to behavior: evolutionary psychology as the missing link. In: J. Dupre (Ed.), *The Latest on the Best: Essays on Evolution and Optimality* (pp. 277–306). Cambridge, MA: MIT Press.
Tooby, J. & Cosmides, L. (1990a). On the universality of human nature and the uniqueness of the individual: the role of genetics in adaptation. *Journal of Personality, 58*: 17–67.

Tooby, J., & Cosmides, L. (1990b). The past explains the present. *Ethology and Sociobiology, 11*: 375–424.

Tooby, J., & Cosmides, L. (1992). The psychological foundation of culture. In: J. Barkow, L. Cosmides, & J. Tooby (Eds.), *The Adapted Mind* (pp. 19–135). New York: Oxford University Press.

Trivers, R. (1971). The evolution of reciprocal altruism. *Quarterly Review of Biology, 46*: 35–37.

Trivers, R. (1974). Parent–offspring conflict. *American Zoologist, 14*: 249–264.

Trivers, R. (1976). Foreword. In: R. Dawkins, *The Selfish Gene*. New York: Oxford University Press.

Trivers, R. (1985). *Social Evolution*. Menlo Park, CA: Benjamin/Cummings.

Waddington, C. (1969). Paradigm for an evolutionary process. In: C. Waddington (Ed.), *Towards a Theoretical Biology, Vol. 2: Sketches*. Edinburgh: Edinburgh University Press.

Ward, P. (1994). *The End of Evolution*. New York: Bantam.

Weissman, A. (1893). *The Germ Plasm Theory: A Theory of Heredity* (English trans.). London: Scott.

Wesson, R. (1994). *Beyond Natural Selection*. Cambridge, MA: MIT Press.

Williams, G. (1966). *Adaptation and Natural Selection: A Critique of Some Current Evolutionary Thought*. Princeton, NJ: Princeton University Press.

Williams, G. (1985). A defense of reductionism in evolutionary biology. *Oxford Surveys in Evolutionary Biology, 2*: 1–27.

Williams, G., & Nesse, R. (1991). The dawn of Darwinian medicine. *The Quarterly Review of Biology, 66*: 1–22.

Wilson, A., & Gedo, J. (1993). Hierarchal concepts in psychoanalysis. In: A. Wilson & J. Gedo (Eds.), *Hierarchal Concepts in Psychoanalysis* (pp. 311–324). New York: The Guilford Press.

Wright, R. (1994). *The Moral Animal*. New York: Pantheon.

Wright, S. (1931). Evolution in Mendelian populations. *Genetics, 16*: 97.

Wright, S. (1932). The roles of mutation, inbreeding, crossbreeding and selection in evolution. *Proceedings of the XI. International Congress of Genetics, 1*: 356–366.

INDEX

Abbott, E., 38, 83
adaptation, current, 8, 28,
    64–65, 105–107, 110,
    113–115, 119–120,
    124, 133–135, 152–
    156, 179–194, 197
    conscious, 28, 82–83, 90–91
    distal, causes of, 11–12
    deep unconscious, 28, 29
    by the emotion-processing
        mind, 6, 8–9, 17, 90–
        97
    -evoking stimuli: *see* triggers
    proximal causes of, 11
        see also psychoanalysis,
        adaptive viewpoint of
    unconscious, 37, 77, 87–91
adaptationist programme, 5,
        9–11, 17, 18, 103–
        104, 116–149, 150–
        159, 160–167, 168–
        176
    for the emotion-processing
        mind, 5, 116–149,
        150–159, 160–167,
        168–176
affects and affective attune-
    ment, 29, 58, 59
aggression and violence, as
    adaptive issues, 132,
    133, 150–159
Alexander, R., 42, 124, 151
altruism, 41, 42
anxiety: *see* death anxiety;
    frame, secured, anxi-
    eties
architecture, of the mind [of the
    emotion-processing
    mind]: *see* mind,
    architecture of
Australopithecus, 117, 118–121

Badalamenti, A., 4, 16, 56, 57,
    77, 164

Badcock, C., xv, 5, 19, 26, 27, 39, 41, 57, 58
Baldwin, J., 13, 66
Barkow, J., 76, 116
Becker, E., 155
behaviours, emotional, causes of:
　adaptive resources, 59
　contextual, 59
　developmental, 13, 59
　inherited, 13, 59
　proximal, 11, 59–60
　see also triggers
Bickerton, D., 16, 18, 29, 68, 77, 107, 131
biology, 161
　psychoanalysis and, xv, 3, 5, 43–46, 56–57
Bloom, P., 20, 70, 77, 107
Bock, W., 114
boundaries: see frame
brain, selectionism and, xiii, 11, 180
Brown, D., 15, 45, 116, 120, 151, 165
Bruner, J., 132
Burnet, F., xiii, 20

Calvin, W., 6, 116
Campbell, D., 64, 68
causes of behaviour and neuroses: see behaviours, emotional, causes of
Chomsky, N., 77, 107
Claus, C., 19
cognition (cognitive functions), 73, 117–118, 122–124, 130–133
　episodic, 118–119
　mimetic, 122–124
　mythic, 132
　and selectionism, 11, 187, 188
　theoretic, 132
communication:
　laws of, 14
　in psychotherapy,
　　conscious, 61
　　unconscious, 37, 61
communicative approach, 4–5, 27–33, 37, 55, 75
　and evolution, 32–33
　hierarchical approach to psychoanalysis: see hierarchies, and psychoanalysis
communicative psychotherapy: see psychotherapy, communicative
condensation, 89
Corballis, C., 77, 107
Cosmides, L., xiv, 5, 6, 7, 10, 11, 12, 15, 18, 45, 56, 58, 59, 63, 64, 66, 68, 69, 76, 114, 187
Cronin, H., 76
cure: see psychotherapy, cure in

Darwin, C., viii, xi, xii–xvi, 1, 5–7, 9–12, 15, 18–28, 33–41, 62–65, 70, 71, 97, 103, 113, 177, 179, 184, 185, 187–190, 195, 202, 203
　evolution, theory of: see evolution, Darwin's theory of
　and Freud: see Freud, S., and Darwin; Freud, S., and Darwinian theory
Darwinism:
　mental, 6, 11, 17–18, 176, 179–194
　universal (Darwin machines), xiii, 70–71, 179–180
Darwin machines: see Darwinism, universal
Dawkins, R., 6, 17, 42, 51, 64, 70, 179

death anxiety (human mortality), 99, 113, 131, 133, 151–159, 166, 174
decoding trigger: *see* trigger-decoding
de Duve, C., 64, 67, 116, 126, 161, 164
denial, 95, 99, 121, 122, 142, 143, 145, 152–157, 171
Dennett, D., 5, 6, 10–12, 20, 23, 26, 34, 63, 64, 66–68, 114, 137, 172
derivative message: *see* messages, encoded
displacement, 85, 170
Donald, M., xiv, 18, 73, 76, 107, 116, 117, 130
dreams, 79–81

Edelman, G, xiii–xv, 7, 11, 180
Eigen, M., 12, 64, 67
Einstein, A., xiv
Eldredge, N., xiv, 6, 13, 17, 20, 42, 51, 52, 63, 114, 189
emotion-processing mind: *see* mind, emotion-processing
empowered psychotherapy: *see* psychotherapy, empowered
encoded message: *see* message, encoded
environment, 12, 184, 185, 192, 193
  expectable, 13–14
  evolutionary (as selection pressure), 10, 12, 44–45, 67, 69, 111, 119–120, 134–135
  as sources of stimuli, 8
  uncertain (uncertain futures), 13–14
  see also evolution, selection pressures

evolution, viii, ix, xi–xvi, 64–72, 76, 110–113, 116–149, 150–159, 160–167, 168–176
  as a constraint on psychoanalysis, 7, 10, 13
  and current adaptations: *see* adaptation, current
  Darwin's theory of (principles of), xii–xiii, 7, 10–17, 20–25
  power of, 7, 21
  epistemology (heuristics; evolution as knowledge acquisition), 63–64, 68–71, 105, 111
    see also evolution, knowledge reduction in
  extinction and, 114, 134, 146
  fitness, 65–68
  Freud and: *see* Freud, S., and evolution; Freud, S., and Darwinian theory; Freud, S., and Lamarckian theory
  as the fundamental theory of biology, 6, 20, 63–71
  historical scenarios in: *see* adaptationist programmes
  Homo sapiens sapiens and: *see Homo sapiens sapiens*, evolution and
  knowledge reduction in, 96–97, 108, 144–147, 153–156
  natural selection, xiii, 5, 10, 40, 64, 65–66, 70, 111–112, 114, 119–120, 168
  unit of (selection), 5, 66, 72–77 [emotion-processing mind as (unit of selection), 72–77]
  see also selectionism

218   INDEX

evolution (continued)
  neo-Darwinian theories of,
    xiii, 63–71
  principles of, 63–71
  psychoanalysis and (evolutionary psychoanalysis),
    xv, 3–18, 41–43, 186–187
  psychology and (evolutionary psychology), xiv, 5
  of repression, 42
  reproduction, favourable (differential) in, xiii, 63, 64, 71
  selection in: see evolution, natural selection in
  selection pressures, 12, 13, 39, 65–68, 111–114, 119–121, 124, 133–135, 136–138
  sexual selection in, 40, 65
  speciation in, xii
  survival and, 39–40
  test in, 63, 70, 110
  variation in, xiii, 63, 65, 70, 110
    see also phenotype; genotype
experience:
  conscious, 82–83, 87–90
  unconscious, 84–85

Fagan, B., 116
frame (ground rules, regulations, boundaries), 30–31, 59, 78–91, 94, 97–99, 160–167, 170–171, 204
  altered: see frame, deviant
  conscious response to, 82–83,
  deviant (altered, modified), 31, 78–83, 88–91, 97–99, 162
    conscious system and, 162, 170, 204
    deep unconscious system and, 162–163, 171
  evolution of, 160–167
  functions of, 163–164
  modifications of: see frame, deviant
  privacy, 79–83
  rectification of (deviant), 85, 90
    models of, from patient, 85, 90
  referrals, from patients, 79–83
  secured, 31, 81, 166, 171, 204
    anxieties, 166 [see also death anxiety]
  setting, 79–83
  time of session, 79–83
  as triggers, 85
  unconscious response to, 84–85, 94
Freud, S., viii, xiv, xv, 1, 14, 19, 20–28, 32–46, 53, 74–76, 84, 186, 187, 198, 199
  and Darwin, 19–25
  and Darwinian theory, xiv, 26–28, 34–41
  and evolution, 4, 19, 26–28, 34–47
    ideas borrowed from, 35–41
  and Lamarckian theory (Lamarckism), xiv, 26, 34, 40–41
  and unconscious domain, 74–77

Gazzaniga, M., 7, 11, 20, 70, 179, 184, 187, 191
Gedo, J., 52, 58, 77
genetic causes of behaviour, 13, 186–187
genotype, 65
Gill, M., 7, 37, 53, 58

Glantz, K., 5, 57
Gould, S., 5, 6, 10, 51, 64, 114
Grene, M., 17, 51
Griffin, D., 165
Grossman, W., 53
ground rukes: *see* frame

Haeckel, E., 19, 34
Hartmann, H., 7, 14, 186
hermeneutics, in psychoanalysis, xv, 22, 44
hierarchies, 51–62, 71
   and biological species, 16–17
   control, 55
   and evolutionary theory, 16, 71, 111–112
   and psychoanalysis, 4–5, 51–62
   structural (nested), 54–55
Homo erectus, 117, 122–128
Homo habilis, 117, 122
Homo sapiens sapiens, 3, 117, 129–149, 150–159, 160–167, 168–176
   emergent features of, 16, 42, 66, 68, 113
   emotion-processing mind of, 29, 98, 129–149, 150–159, 160–167, 168–176, 179194, 195–207
     see also minds, emotion-processing
   evolution and, 11, 12, 66, 114–115, 129–149, 150–159, 160–167, 168–176
   shared features of (with other species), 16

id: *see* psychoanalysis, structural theory of
immune system, xiii, 11, 184
incest, 151, 165
individuality, 14–16, 45

instructionism, 11, 22, 112, 179, 180, 181, 182–184

Jacob, F., xi
Jerne, N., 11, 20, 70, 187
Johanson, D., 116

Kauffman, S., 67, 181
Kitcher, P., 3, 4, 5, 19, 27, 34, 35, 56, 59
knowledge acquisition, 9
Kriegman, D., xv, 3, 4, 5, 6, 10, 11, 13–15, 26, 27, 37, 40–42, 57–59, 73, 186
Kris, E., 14, 186
Kuhn, T., 55
Kuper, A., 116

Lamarck, J., xi–xiv, 11, 26, 34, 35, 38–41, 182
Lamarckism, xi–xii, xiv, 11, 26, 34, 35, 38–41, 182–184
   and Freud: *see* Freud, S., and Lamarckian theory
Langs, R., vii–ix, xv, 4, 12, 15, 16, 28, 29, 31, 41, 42, 55–58, 61, 74, 77, 83, 84, 93, 132, 141, 164, 190, 201, 203
language, 68, 77, 90, 107, 131–134, 137
   see also mind, emotion-processing, language basis for
Leakey, R., 116, 117, 134, 146
Lewin, R., 116, 117, 134, 146
Lewontin, R., 5, 10, 51, 66, 72, 114
Liberman, P., 42, 77, 107, 130
listening–formulating process, 28–29, 81–85
Lloyd, A., xv, 5, 41, 42, 57, 58, 73, 186
Lowenstein, R., 14, 186

Malthus, T., 23
meaning
　encoded, 29
　manifest, 28–29
Mayr, E., 5, 6, 10, 114
Mendel, G., 34
message, encoded, 147–149
Miller, G., 125
Millikan, R., xiv, xv
mind, emotion-processing, 5, 7, 8–10, 29–30, 72–99, 104–110, 120-122, 124–128, 134–149, 150–159, 160–167, 168–176, 187–194
　architecture (design, structure) of (psychoanatomy), 15–16, 58, 86–87, 91–99, 105–107, 120–122, 124–128, 134–135, 137–149, 187–194
　conscious system of: see system, conscious
　as a Darwin machine: see Darwinism, mental
　design of: see mind, emotion-processing, architecture of
　dysfunctions (syndromes) of, 201, 202–204
　evolution of, 105–107, 116–149, 152–159, 187–194
　and *Homo sapiens sapiens*: see *Homo sapiens sapiens*, emotion-processing mind of
　language basis for, 8, 77, 89–90, 107
　laws of, 14
　models of:
　　communicative, 29–31, 90-97
　　structural, 75–76
　　topographic, 74–75
　puzzling features of, 107–110, 168–173, 194
　systems of: see system
　unconscious system of: see system, unconscious; system, deep unconscious
　as unit of selection: see evolution, selection in, unit of, emotion-processing mind as
　and universal Darwinism: see Darwinism, mental
Munz, P., xiv

narrative, 132
　as carriers of encoded meaning, 30, 84–85
natural selection: see evolution, natural selection in; mind, emotion-processing, evolution of
neo-Darwinian theories: see brain, selectionism and
Nesse, R., 5, 11, 15, 39, 41, 42, 56–58, 73, 111, 114, 186, 194
neural Darwinism: see brain, selectionism and

Ornstein, R., 12, 132

Pearce, J., 5, 57
phenotype, xiii, 65, 66, 110–112
Piaget, J., 33, 44
Pinker, S., xiv, 20, 70, 77, 107
Plotkin, H., xiii, 3, 6, 11, 14, 17, 20, 22, 39, 51, 52, 54, 63–70, 73, 110, 114, 173, 175, 179, 184, 191
Popper, K., xiv
psychoanalysis, ix

adaptive viewpoint of, 7, 8, 21, 23–24, 28–32, 35, 45–46, 57
and biology: see biology, psychoanalysis and
and evolution: see evolution, psychoanalysis and
formal science for, 15, 43, 59, 77
and hermeneutics: see hermeneutics, in psychoanalysis
hierarchies and: see hierarchies, and psychoanalysis
intrapsychic fantasies, theory of, 12
metapsychology of, 7, 13–14
natural defences against, 24–25
and science, 3–4, 46, 55–57, 59
structural theory of, 32, 74, 75–76, 187
systems theory for, 57
and unconscious processes, 23, 33
psychoanatomy: see mind, architecure of
psychology, and evolution: see evolution, psychology and
psychotherapy, ix, 61, 182–184, 195–205
communicative, 61
cure in, 199–202
empowered, 61

Rapaport, D., 7
repression, 42, 85, 86, 95, 121, 142, 145
reproduction, favourable: see evolution, reproduction favourable in
Restak, R., 8
Ridley, M., 6, 64, 116

Ritvo, L., 19, 26, 34, 36, 39

Salthe, S., 17, 51, 54
scenarios, evolutionary: see adaptationist programmes
science:
formal (mathematical), for psychoanalysis: see psychoanalysis, formal science for
hierarchies and, 52, 55–57
Searles, H., viii
selection: see evolution, natural selection in
selectionism, xiii, xiv, 11, 22, 112, 179, 180, 181, 184–186, 195–205
dysfunctional (selections), 185–186, 189–190, 196, 198-199, 201, 202–206
self-awareness, 12, 132
self-organizing patterns, 16, 67, 181, 185
Shreeve, J., 116
Shulman, D., 57
Simon, H., 114, 148
Slavin, M., xv, 3–6, 10, 11, 13–15, 26, 27, 37, 40–42, 57–59, 73, 186
Smith, D., xv, 28, 74
speciation: see evolution, speciation in
structural theory: see psychoanalysis, structural theory of
subliminal perception: see perception, unconscious
Sulloway, F., xiv, 4, 16, 19, 24, 26, 34, 35, 39, 43
superego: see psychoanalysis, structural theory of
survival: see system, conscious, survival function of

system:
  conscious, 29–30, 90–97, 107–110, 120–121, 124–128, 136–137, 142–143, 144–145, 153–156, 169, 172–173, 183, 187, 192–193
    adaptations by, 105–106
    defensiveness of, 142–144, 145, 153–156, 170
    denial by, 142, 153–156 [see also denial]
    evolution of, 82, 120–121, 124–128, 136–137, 141, 142–143
    and rules and boundaries, 106
    superficial unconscious, subsystem of, 105
    survival function (strategy) of, 111, 112–113, 125–128
  deep unconscious, 29–30, 90–97, 106–110, 140–142, 144–148, 156–159, 183, 193–194
    adaptive recommendations of, 85–86, 90, 106
    encoded communications from, 106
    fear-guilt subsystem, 93, 106–107, 127–128, 156–159, 170, 194
    processing (adaptations) by, 8
    and rules, frames, and boundaries, 106
    wisdom subsystem (intelligence of), 106, 144–145, 193–194
  overload, 126–127, 136–137

therapy: see psychotherapy

Tooby, J., 5–7, 10–12, 15, 18, 45, 56, 58, 59, 63, 64, 66, 68, 69, 76, 114, 187
topographic model of the mind: see mind, emotion-processing, models of, topographic
transference, 36–37, 83
trigger-decoding, 29, 44, 84–86, 87–91, 174
triggers, 30, 59–61, 84–87, 89–91
  conscious, 82–83, 86–87
  decoding method: see decoding process
  as environmental impingements: 12, 86–87
  frame-related, 30
    see also frame
  nature of, 28
  unconscious (repressed), 84, 86
Trivers, R., 42

unconscious, as a term and concept, 33, 47, 74–76
unconscious experience: see experience, unconscious
unconscious perception: see perception, unconscious
unit of selection: see evolution, natural selection in, unit of
universal Darwinism: see Darwinism, universal; Darwin machines
universality, 14–16, 45, 72, 191

validation, of therapist's interventions, vii, viii
variation: see evolution, variation in

vignette, clinical, 78–91, 97–99, 143–144

violence: *see* aggression and violence, as adaptive issues

Waddington, C., 14, 69

Ward, P., 116, 134, 146

Weissman, A., 34

Williams, G., 5, 11, 41, 57, 63, 111, 114, 194

Wilson, A., 52, 58, 77

Winnicott, D., vii

Wright, S., 12, 64, 76